DAIRY DEVELOPMENT IN THE NEW MILLENNIUM

The Second White Revolution

DAIRY DEVELOPMENT IN THE NEW MILLENNIUM

The Second White Revolution

DR. MOHAN PD. SHRIVASTAVA
University Professor
Department of Economics
Magadh University,
Bodh-Gaya, (Bihar)

and

DR. JAISHANKAR SINGH
Assistant Commissioner
Commercial Taxes, (VAT)
Jharkhand

DEEP & DEEP PUBLICATIONS PVT. LTD.
F-159, Rajouri Garden, New Delhi - 110 027

DAIRY DEVELOPMENT IN THE NEW MILLENNIUM
The Second White Revolution

ISBN 978-81-8450-102-5

Typeset by RAHUL COMPOSERS
358, Pocket-B, Phase-II, Sector 16-B, Dwarka, New Delhi - 110 075

Printed in India at MAYUR ENTERPRISES
WZ Plot No. 3, Gujjar Market, Tihar Village, New Delhi - 110 018

Published by DEEP & DEEP PUBLICATIONS PVT. LTD.
F-159, Rajouri Garden, New Delhi - 110 027 • Phone : 25435369, 25440916
E-mail : ddpubs@gmail.com • ddpbooks@yahoo.co.in
Showroom :
2/13, Ansari Road, Daryaganj, New Delhi - 110 002 • Telefax : 23245122

Contents

Preface

The Indian dairy sector, with an annual production of 96 million metric tones, is the largest in the world. Currently, dairy is the India's second largest food expenditure category, after cereals, accounting of 18 per cent of the total food expenditure.

During the last three decades, Indian agriculture has undergone a huge transformation, mainly through technological innovations, resulting in Green, White, Blue, Yellow, rather Rainbow Revolution in different aspects of food production.

It is a known fact that in India "White Revolution" became a reality greatly due to the dairy cooperative movement pioneered initially by AMUL (Anand Milk Union Ltd. and through the subsequent efforts made to replicate AMUL, what came to be famously known as "ANAND PATTERN" in some other parts of the country. This ultimately led to the setting up of National Dairy Development Board (NDDB) with its headquarters at Anand, Gujarat. It is not just a coincidence that free India's two great personalities viz. the first Union Home Minister Late Sardar Vallabh Bhai Patel and the Second Prime Minister, Sri Lal Bahadur Shastri, provided the original concepts for Amul and the NDDB.

Today, the exports of dairy products from India are increasing. The Asian countries are the major destination of our exports. In recent years, India has found some new markets in South and East Asia and in the gulf regions. China, South and North Korea, Lebanan, Myanmar, Israel and Iran are the new and emerging markets. Considering the size of these markets, particularly of China, India has a great exports potential in these markets. Skim Milk Powder (SMP), Whole Milk Powder (WMP), Milk Food for Babies, Other Milk Powder, Butter, and Ghee are the main products of India basket of dairy exports.

India needs to diversify her product range to capture the new markets as the food habit pattern is changing rapidly in South and East Asian markets. Although, India has made some positive efforts to diversify her exports towards products like Whey-based products, Dairy Spread, Butter Oil, Processed and Powdered Chees and fermented products, but further efforts need to be initiated and intensified in this direction. The export potential in neighbouring countries in India subcontinent and in the Gulf region however, still lies in traditional products. Nepal, Saudi Arabia, Oman and UAE will be the major market for concentrated milk and cream-based products and butter or fat-based products.

India with its enormous indigenous bovine population and indigenous method of agriculture has comparative advantage in converting to organic farming, to meet the increasing demand for safe food including organic food products locally and internationally. The Indian dairy farmers who are contributing towards India's number one position in milk production, which depends on certain rules, procedures and certification standards to be observed. A farmer who would like to convert from conventional milk production to organic livestock farming standards. Moreover, with increasing concern for environment and rising consumer awareness about safe and quality foods, the organic foods are attracting ever-increasing number of consumers. However, the demand for organic foods and products has created new export opportunities for the development countries like India.

The present growth trends of around 9 percent as stated by Central Statistical Organisation of India, The Economic Survey, 2006-07 and the Union Budget, 2007-08 has catapulated the nation's aspiration to attain the status of a developed country, much ahead of the other emerging nations.

While aiming at faster growth the Approach Paper to the Eleventh Plan also emphasises the need for making growth more inclusive and recognises that this is in many ways more difficult.

Thus, dairying must be helpful in removing poverty and unemployment. There is an urgent need for organised dairy marketing from stage of production to marketing and profitable utilisation of dairy products. Modern management and technical process in relation to production, processing, transportation, promotion and distribution, cost-benefit analysis, price incentives, technical inputs, etc. need fresh evaluation and indepth micro-explanation.

The growing competitiveness, triggered by WTO, has made the global dairy markets increasingly complex. Each year, a large number of new food products are being added in the market place in response to the consumer's search for delicacies that are novel but natural with touch of mystery and class.

Since initiation of economic reform impressive advances have been made in mechanised production of wide range of indigenous milk products. Our traditional dairy products sector is poised for rapid expansion with the result of application of modern process technology in the production of indigenous sweets. The rising demand for packaged fresh products like *dahi, lassi, paneer* and *cream butter* is widening the base of modern dairy sector.

Singnificantly, technological innovations have come at time when ethnic milk products are attracting worldwide attention. Markets for traditional Indian products also exist overseas where ethnic population of Indian sub-continent has settled. Moreover, to tap these markets, it is essential to confirm to the globally standards of food quality and safety. It calls for Good Manufacturing Practices (GMP), ISO Quality Management System (QMS), Hazard Analysis Critical Control Points (HACCP) and food standards by FAO/WHO Codex Allimentarious Commission. With growing consumer awareness towards health and nutrition packaging and nutritional labelling have become important. This trend has been further accelerated by the changing dietary habits and lifestyle to the ever increasing number of consumers. They are demanding convenient, easy to cook, ready to eat food stuffs in appropriate packaging that return freshness, flavour and taste, preserves nutrition and also a long shelf-life.

Consequent upon this, all related issues of Second White Revolution in the form of dairy development in the new millennium have been explained in the book in a quite systematic, simple and lucid manner. It is hoped that the book will fill up the information gap and will be useful to all concerned.

However, we invite healthy and constructive criticisms and the suggestions from the readers for the value and quality addition to the book.

MOHAN PD. SHRIVASTAVA
JAISHANKAR SINGH

1

Introduction

Dairy farming or dairy industry is an important occupation of Indian farmers because dairy food forms and important constituent of dairy menu for the proper world. It is capable of providing/producing enormous income and employment by way of rearing milk animal on scientific lines. We know, India is a vast country with lot of diversity in the food habits of her population. However, foods invariably form a pride of place in the diet of Indian population irrespective of regional and cultural differences (*Indian Dairyman*, 55, 6, 2003, p. 68).

A significant phenomena of our times is the "*White Revolution*"—India's contribution to co-operative dairying. The "*Operation Flood Programme*" has caught the imagination of planners the world over Dairy India, 1992, p. XI).

Operation Flood proposed by the National Dairy Development Board (NDDB) has restructured the milk markets in India and its growth. Operation Flood-I laid the foundation of modern dairy Industry in India. There was a viable self-sustaining growth of produce controlled by dairy co-operatives. The Operation Flood-II and Operation Flood-III that followed Operation Flood-I showed commendable achievements in the dairy sector. "As per World Bank expert opinion, for an initial investment of Rs. 200 crores in Operation Flood-II, the not return/ year to rural economy had been 2400 crore rupees. No other major development programme all over the world has matched this input-output ratio"—(*Kurukshetra,* 2006—Volume 52, No. 2, p. 31).

Today, as President of India, while addressing joint session of both the houses of Bihar Legislative Council and Assembly on the 28th March 2006 for the fist time in Parliamentary democratic system, clearly stated that we need both *second green revolution* as well as *second wave of white revolution* for rapid and sustainable growth of Bihar and the nation as a whole. It can become a panacea to alleviate poverty, massive increase of employment opportunities in rural areas, remove regional imbalances enhance capital formation, export earning; per capita income and reduce fiscal and revenue deficits.

India is the world's single largest milk producing country with a share of about 14% in World milk production. Milk has achieved a unique status in terms of its output value and contribution to the national economy, with output value exceeding Rs. 1,00,000 interim of number of milk produces and quantity of milk produced. Over the last two decades, while both population and food grain production grew at around 2%, milk production grew at more than double the rate of growth of the population, increasing per capita availability of milk from 112 grams/day in 1970-71 to 231 gms/day in 2003-04 (*Kurukshetra,* Volume 52, No. 2, p. 31). Dairy development in India has been acclaimed world over as one of modern India's most accomplished development programmes. The states like Gujarat, Maharashtra, Uttar Pradesh, Haryana, Rajasthan, Andhra Pradesh, Karnataka and Tamil Nadu are surplus in milk production. Productivity performance of dairy Industry across the country has registered an annual growth of 17.1% for the country as a whole with a highest market share of 23.5% for Maharashtra followed by Gujarat (17.2) (*The Hindu Survey of Indian Agriculture,* 2004, p. 110.

The most needed linkage between produces and the consumers had been established resulting in a '*white revolution*'. Policy linkaged involved pricing structure, procurement and marketing, system, breeding, cattler food diseased control and veterinary services. The launching of technology mission by Government of India in 1987 is another major milestone for dairy development in the country

Since ancient times, man has domesticated livestock, partly for his survival against starvation and partly for having nutritious and tasty animals foods as part of his dietary menu. According to Dr. V.M. Rao, "Archaeological *finds* from the Indus Valley Civilization *attest* to the domestication of both cows and buffaloes establishment of cattle breeding farms. On the recommendations of Royal Commission on Agriculture, 1928 for, the cattle development formed an important part of the earliest developmental activities in India. Another important contribution during the late 1930's and early 1940's in the field of cattle development was the laying down of description and characteristics of different breeds of cattle and buffaloes by the Imperial Council of Agricultural Research (ICAR). There were 904 veterinary hospitals and dispensaries in British India at the time of submission of Report of the RCA." (V.M. Rao, *Dairy Farming*; Reliance Publication, Delhi, 1991, p. 3)

But in 1930's, Wright (1937), while emphasising the importance of improving the milk yield in Indian cattle, strongly opposed adoption of large cross-breeding to improve the milk yield of local cattle under the prevailing local conditions and recommended that immediate steps should be taken in selecting and improving milk strains of indigenous cattle. Further, Animal Husbandry wing of the Board of Agriculture and Animal Husbandry of India at its Fourth (1940) and Sixth (1945) (*NCUL*, 1982) meetings discussed the question of mass improvement of inferior indigenous cattle and recommended that grading up of cattle should be taken up in a systematic way in selected areas. In 1944 the Government of India invited Papperall for making a rapid survey of the problems facing the dairy industry in the country. (R.A. Papperall (1945) *The Dairying Industry of India : Report as an investigation with the recommendations;* Govt. of India, Shimla).

The domestication of cattle occurred between 6000 and 10000 years ago. According to Henry Eckles and W.B. Ombas and Harold Macy (18th Ed., 2004), "Not much is known about the history of this period, but men probably hunted cattle and wild animals prior to the time that they were domesticated. In ancient

India, the oldest written records of the human race are found in the Sanskrit. These records date back nearly 6,000 years, but milk had already become an important article of food. Infact, the cow was so important to these early peoples of Central Asia that wealth was measured in number of cattle. Later the cow was made a sacred animal and still so considered by a part of the population of India.

The cow was also worshipped in Basylonia and Egypt about 2000 B.C. Hathor, the goddess who watched over the fertility of land was depicted as a cow. Over 50 references to cows and milk are found in old Testament and promised land was desired as *"a land flowing with milk and honey."*

The soldiers of Chenghis Khan, the Mughal emperor who conquered Asia and a large part of Europe in the thirteenth century, earned dried milk as a part of their ration. Cheese was an important part of the food supply carried by the Vikings in their Voyages which carried them to the shores of all northern Europe and even across the Atlantic.

It was further stated and referred to by these, above authors that "From these early days to present, the cow has continued to be the servant of man and her importance as a source of food has not been diminished by thousands of years which have passed. The then Governor ordered that one cow and two goats should be brought over for each six people".

The Dairy Industry prior to 1850 was also in progress. According to Eckles, Comb and Macy in their book *Milk & Milk Product* (published by Tata McGraw Hill, New Delhi, 2004, p. 3) "For over 225 years following the first settlements in America very little change took place in the method of producing milk or in the manufacturer of dairy products. Most of the population lived on farms or in small villages, and each family dependent upon its own cows for the family milk supply. Nearly all the cows freshed in the spring or early summer, and an abundance of milk was available as long as pastures were good. Many of the cows completed their location period in the fall and very little milk was available for human consumption during the winter months. Often times cattle became so week from lack of feed during the winter that they had to be helped to their feet and sometimes, actually starved to death.

During the summer the surplus milk was manufactured generally into butter or cheeses on the farms having several cows in milk. The market prices for these products during the summer season if any market were available at all, were ridiculously low. In the winter, fresh butter was not available on the market at any price. The surplus butter made in summer was packed largely in firkins and covered with brine. It was stored in cellars or spring houses and constituted the chief source of butter for winter use. People were accustomed to using butter with a more or less rancid and old flavour. The product sold for table use 100 years ago would be rejected entirely at present extent for industrial uses.

During that period prior to 1850, cheese manufactured in the home was generally far superior in quality to the butter since cheese was easier to keep under the rather primitive storage conditions. Probably much of cheese represented a grade equal to that of cheese now on our markets. Between 1830 to 1850 certain farming sections in Vermont and New York began to develop as centres of production for surplus butter and cheese. These products were marketed in the rapidly expanding cities of the Northeast. Gradually farmers within easy driving distance began delivering milk over regular routes in the cities. This was the beginning of the fluid milk sheds which surrounds our large cities today. Prior to 1850, however, most milk was necessity produced within a few miles of where it was consumed because of the lack of suitable means of transportation and refrigeration. Generally speaking, prior to 1850 was characterised by gradual growth in the dairy industry. The development was not spectacular, but the stage was set for the many changes which were to take place in the years to follow.

THE WHITE REVOLUTION/DAIRY INDUSTRY AFTER 1850

According to Eckles Comb and Macy (2004), "The middle of nineteenth century is a convenient point to set as marketing the beginning of modern dairying. More changes have been made during the past 100 years than in all previous 60 to 100 centuries since cattle was domesticated. When one observes a modern creamery, ice-cream factory, fluid milk bolting plant, condensary or dry milk plant, he should appreciate that the development of these modern factories has occurred, to a very large degree, within the past 30 years."

Following factors played most important role in the evolution of White Revolution in the following form of :

1. The Factory System
2. Improved Machinery
3. Transportation
4. Economic Factors
5. Research and Scientific Investigation
6. Improved Livestock
7. Milk Production and Utilisation

Organisation and maintenance of Goushalas (house for old and unproductive cows) was another important effort made in India during pre-planned period. For the last two centuries in the country, they are being maintained on the religious and economic considerations. Government of India set-up *"Central Goushala Development Board"* to develop Goushalas as Centres for cattle breeding and milk production. Later, the *Central Council of Gosamvardham* (CCG) was contributed in 1952 to function as a co-ordinatory advisory body on cattle development. However, Goushalas did not make any significant impact either on cattle development or on milk production in the country. Consequently, both central and state governments did not show much importance on the development of Goushalas and finally the CCG was closed by 1969.

For the first time in 1920, *"An Imperial Dairy Expert"* was appointed for organising training programme on the recommendation of the *Board of Agriculture* of the Government of India. Subsequently, Imperial Institute of Animal Husbandry and Dairying was also established in 1923 at Bangalore to co-ordinate the efforts of the Government.

A beginning was made in 1946 for the first time in India, milk produced in rural areas of Khaira district was collected in bulk pasteurised and transported by rail for distribution in Bombay metropolis. Similarly, pasteurised and bottled milk was set-up for sale in India for use of common man on December 15, 1950 from the Aarcy Dairy (Bombay).

WHITE REVOLUTION DURING PLAN PERIOD

According to V.M. Rao, Dairy Development in India received a fillip after independence when industrialisation and public awakening warranted the establishment of organized milk collection, processing and distribution of milk to cater to the urban needs.

In 1955, the *Indian Dairy Research Institute* was transformed as the National Dairy Research Institute (NDRI) and was shifted from Bangalore to Karnal. During Second Five Year Plan (1956-61), 7 *liquid milk plants* were completed, 8 *pilot milk schemes* as precursors to the establishment or main-dairy plants were started together with establishment of three milk creameries and 02 milk products factories, 4 cattle salvage farms were also started.

In the Third Five Year Plan (1961-66) provision of Rs. 37 crores comprising of Rs. 32 crores for the state schemes and Rs. 4 crores for the central schemes were made. The *National Dairy Development Board* (NDDB) was set-up by the Government of India in 1965 at *Anand* (Gujarat) with the major objectives or providing, on a non-profit basis and technical services in building-up dairy projects.

During Fourth Five Year Plan, all these important programmes continued. The most significant central sector scheme in Fourth Plan and being continued to date relating to dairy development has been the outcome of an agreement signed in March, 1970 between Government of India and the World Food Programme, popularly known as 'Operation Flood Programme (OFD)', for the supply of 26 lakh tonnes of skim milk power and 0-42 lakh tonnes butter oil.

The Indian Dairy Development Corporation which has been set-up in pursuance of this agreement was expected to receive, store and sell these donated commodities to public sector milk plants and generate funds of the order of about Rs. 95 crores over a period of five years.

These funds were to be utilised for expanding the exiting capacities of dairies at Delhi, Madras, Calcutta and Bombay and for undertaking programmes for increasing milk production in the hinter land which comprises of areas in 10 states and Delhi. The scheme went into operation in July, 1970 and ended on 31st March, 1981. Admittedly, the first phase has taken a much longer time than originally contemplated to achieve its goals. The capacities of four metro dairies under this project were enlarge from 10 lakh liters to 31 lakh liters per day.

The major programme in the dairy sector in Sixth Plan was *Operation Flood Programme–II* (OFP-II) which covered 21 states and 4 union territories. This project was being implemented through a three tie co-operative structure with state federation at the *apex level* with unions at district level and primaries at the village level. Under the project, a *national milk grid* was developed to cater to be milk supply of the four metropolitan cities and all towns having a population of one lakh and above. At the end of the Sixth Plan, there were 166 *liquid milk plants*. Further, three centrally sponsored dairy development project were implemented in many districts to develop areas not covered in *Operation Flood Programme-II*.

Operation Flood Programme-II continued in the Seventh Plan with an effective impact on the production and marketing of dairy farming. In 1987 a *Technology Mission* on dairying was established to : (i) reduce cost of operation, and (ii) ensure greater availability of milk and milk products.

In 1986 *Operation Flood-III* was proposed by the Government of India to EEC and World Bank to further extend and consolidate. The dairy development in the country by strengthening. The grass-roots level infrastructure facilities for milk procurement, milk processing and marketing network established under *Operation Flood-I & II*. The *Operation Flood-III* was proposed upto 1994 which is mainly intended to cover imbalances remaining after *Operation Flood-I & II* with a coverage to additional 15,000 villages and 200 medium size urban centres. The Operation Flood Programmes thus, have been implemented with the aids and grants received from *World Food Programme* (WFP) of the *United Nations Food and Agricultural Organisation* (FAO), the *European Economic Community* (EEC) and the *World Bank*.

Operation Flood-II & III have involved an outlay of Rs. 4855 million and Rs. 6,806 million respectively. Out of total funds received about one-third of the total cost of these projects have been funded by World Bank and rest about 50%, originated in the form of dairy community, aid from *European Economic Community*.

White Revolution/Dairying further received top priority in the 8th & 9th plans. The contribution of animal husbandry and dairying total gross domestic product (GDP) was 19 percent in 2000-2001 at current prices. The contribution of the milk group alone was Rs. 1,01,990 crore which was higher than wheat (Rs. 47,091 crores) and sugarcane (Rs. 27,647 crore). According to Ninth Plan milk production in India remained more or less stagnant from 1950-70. Thereafter, it increased rapidly, reaching 84.6 million tonnes (mt.) in 2001-02. But the Ninth Plan target of milk production (96.49 mt.) was not achieved. The per capita availability of milk increased from 112 gm. per day in 1973-74 to about 226 gm. per day in 2000-01. However, it was still below the world average of 285 gm. per day. Investment in the dairy sector in the Ninth Plan decreased significantly compared to the Eighth Plan out of 168 milk unions, 58 milk unions (34.5 per cent) were running in loss as of March 2000.

So far the Government policy in the dairy sector has been to give preference to the establishment of milk processing plants linking rural milk producers to urban consumers through a network of co-operatives. Restrictions on establishing new milk processing capacity under Milk and Milk Products Order (MMPO) has now been removed. No policy measures have been undertaken so far to give a fillip to the unorganized sector involved in the production of Indian Dairy Products like Ghee, Paneer, Chhena, Khoa, etc. which have tremendous potential in the export market in Asia and African countries.

In *Tenth Five Year Plan* dairying with animal husbandry has further been receiving high priority in the efforts for generating wealth and employment increasing the availability of animal protein in the food basket and for generating exportable surpluses. The overall focus is on four broad pillars viz.:

(i) removing policy distortions that is hindering the natural growth of livestock production,
(ii) building participatory institutions of collective action for small-scale farmers that allow them to get vertically integrated with livestock processions and input suppliers,
(iii) creating an environment in which farmers will increase

investment in ways that will improve productivity in the livestock sector, and

(iv) promoting effective regulatory institution to deal with the threat of environmental and health crisis stemming from livestock.

The *Tenth Plan* target for milk production was set at around 108-4 mt. envisaging an annual growth rate of 6.0 percent. In Tenth Plan following initiatives were taken by the Government :

- Withdrawal of Milk & Milk Product Order (MMPO).
- Introduction of National Project on Cattle & Buffalo.
- Database & Information Network.
- Creation of disease free zone.
- Conservation of threatened livestock breeds.
- Feed & fodder production enhancement.
- Dairy/Poultry Venture capital fund.
- Clean Milk Production.

DAIRY STRATEGY DURING TENTH PLAN (2002-07)

According to the Tenth Plan, 2002-07 "The contribution of animal husbandry and dairying to total gross domestic product (GDP) was 39% in 2002-01 at current process. The contribution of Milk group alone (Rs. 101,990 crore) was higher than wheat (Rs. 47,091 crore). It is estimated (1993-94) that almost 18 million people are employed in the livestock sector in principal (9.8 million) or subsidiary (8.6 million) status. Women constructed about 70% of labour force in live stock farming. The overall growth rate in the livestock is steady (around 4.5%) in spite of fact that investment in this sector is substantial as the ownership of livestock in more evenly distributed landless labourers and marginal farmers, the progress in this sector will result in a more balanced development of the rural economy.

The broad framework of the cattle and buffalo breeding policy being followed since the mid-sixties envisaged selective breading of indigenous bread in their breading tracts and use of such improved breads for upgrading of the non-descript stock.

The present production capacity of frozen seaman dosses is about 30 million as against the estimated requirements of 65 million dosses annually.

MILK PRODUCTION

Milk production in India remained more or less stagnant from 1950-70. Thereafter it increases rapidly, reaching 84.6 million tonnes in 2001-02. But the Ninth Plan target of milk production 96.49 mt. was not achieved.

The per capita availability of milk increase from 112 gm per day in 1973-74 to about 226 gm in per day in 2001-02. However, it is below the world average of 285 gm per day investment in dairy sector in the 9th Plan decreased significantly compared to 8th Plan out of 168 Milk Unions, 58 Milk Unions were running in loss till March 2000. But no policy major have been undertaken so far to give a Philip to the unorganised the sector involved in the production of dairy product which have tremendous potential sports market in Asian and African Country.

Under 11th Plan annual husbandry dairy has received high priority for generating wealth and employment and exportable surplus. The 11th Plan targeted milk production at 108.4 mt. envisaging an annual growth rate of 6%. Use of technological and marketing intervention in production, processing and distribution of milk products have been made central theme of any future programme. 11th Plan also laid emphasis that the bacteriological quality of raw milk at the time of milking in India is comparable with that in the advanced designation to some extent holistic approach would be taken to address the issue of clean milk production, which is imperative for making and promoting export of dairy products. Steps would be taken for development of unorganised milk sector that control a significant portion of liquid milk and sweet meat market (10th Plan, Vol . III, pp. 567-72.

During Eleventh Plan so far as milk production is concerned, "the bacteriological quality of raw milk at the time of milking in India is comparable with that in the advanced dairying nations. Subsequently, however, the quality deteriorates due to improper handing of milk and lack of availability of infrastructure like all weather roads, cooking facilities, potable water, regular electric supply and sewage disposes. A holistic approach is being taken to address the issue of clean milk production, which is imperative for marketing and promoting export of dairy products. Steps are also being taken for the development of unorganised milk sector that controls a significant portion of the liquid milk and sweetmeat market.

India through systematic planning in forms of various dairy development programmes has reached the present milk production level of 20 million tonnes topping the world in terms of milk products.

Due to co-operative movement Indian Dairy industry has become producer-oriented. The present achievement is the fruitful cumulation of the sustained systemic and intensive afford made by government of India and Bihar as well through various programmes.

CHALLENGES FACING INDIAN DAIRY

The present dairy sector in India has been facing various challenges despite India has made remarkable strides in the arena of dairy development, i.e.

(i) Socio-economic challenges;
(ii) research and extension;
(iii) milk procurement and processing; and
(iv) organized challenges.

The cross-breeding technology has helped in increasing the milk production, though the extent of contribution of cross-breeds to this is not clear and the varies across India. Though the impact of cross-breeding programme on milk production is not as substantial an expected, there is no point in discounting it. It is better to implement the cross-breeding programmes with vigour only in the potential promising areas rather than wasting resources in implementing the programme throughout the country. Important conditions influencing the successful implementation of cross-breeding include an increased demand for cow milk or young stock resulting in a remunerative market, low demand for drought animals, accessibility of veterinary assistance to farmers and low prices of high quality feeds relative to milk prices.

MILK AND DAIRY PRODUCTS AS FOOD

For centuries, milk has been recognized as an almost indispensable food for mankind. Its value in the human diet and as a supplementary feed for animals was known for hundreds of years before it came under the scrutiny of the science and technology of chemistry. The value of milk based upon more deeply seated fundamental facts than merely the amount of protein and energy it supplies to the body. It supplies high quality proteins in the human diet. Milk also supplies liberal amounts of calcium which is often otherwise deficient in the human diet. It is also excellent source of the vitamins necessary to health. The food requirements include followings : (a) water, (b) carbohydrates, and fat, (c) protein, (d) mineral matter, and (e) vitamins (accessories to growth and health).

CONSTITUENTS OF MILK

Milk as an article of food for mankind antedates the earliest recorded history. Milk may be defined as the normal secretion of the mammary gland of mammals. Nature designed milk as a food for the young. Mankind, thousands of years ego, learned the possibilities of milk and milk products as a food not only for the young but for adults. Although, all the species mentioned are used as a source of milk by certain people, the cow supplies such a large proportion of the product, used especially in Europe and America, the little attention need be directed towards the product of other species. The term 'milk' always will be understood as referring to the milk of the cow unless other species are mentioned specifically.

MAIN CHARACTERISTICS OF MILK

Milk contains following main characteristics :

(a) Milk ranges in colour from a bluish-white to an short golden yellow, depending upon the cattle the amount of fat and solids.

(b) In the large qualities, present milk appears entirely opaque, while in thin layers it is somewhat transparent.

(c) Milk has no pronounced taste but is slightly sweet to most person.
(d) Fresh milk is tested with litmus.
(e) Milk appears unchanged by seating until a point near the boiling point is reached when a tough film forms on the surface.

Milk contains major constituents like water, protein, sugar, fat and ash. It also contains a number of constituents in addition to above, which are of minor importance particularly with regard to the proportions in which they occur in milk. Strengths, Weaknesses, Opportunities and Threats (SWOT) of dairy development have become a serious subject matter of debate during these days which plays a paramount role in understanding the management problems at all stages irrespective of various type of organisations and programmes. Successful dairy development requires the ability and capacity on the part of planners to make correct decisions.

To improve the management of resources and to achieve the desired level of success in the dairy development process, the dimensions such as formulating goals or objectives for development, recognising the problem and opportunity for development, obtaining of information about possible threats by constraints, specifying and analysing the alternatives, successful implementation and evaluation of the programme are more important. They play the decisive role. Lack of relevant materials or information about the development potentials opportunity for diversification, system ability to understand the development process and possible alternatives for development are the severe constraints the decision–making process.

Thus, the relevant information about strengths, weaknesses, opportunities and threats (SWOT) are associated with the dairy development system in our country are of paramount importance.

Strengths refer to the basic assets or potentials of a thing that would provide competitive advantage for its growth and development. *Weaknesses* are the liabilities that can create a state of time and situation which acts on a specific disadvantage for the growth and development. *Opportunities* means the ability to grow and achieve the given specific objectives in a given direction and situation. *Threats* are the situations that block the abilities to grow, develop and progress in a given situation or environment.

(A) Strengths of Dairy Development

Following aspects are the basic strength of our dairy development which assure the livelihood for millions of farm families in rural India :

(a) heterogeneous and well diversified farming locations;
(b) less merchanised agricultural farming;
(c) high demand for milk and milk products;
(d) sound research and technological support;
(e) infrastructure in terms of physical and manpower resources;
(f) establishment of three-tier Anand type of co-operatives and farmers' participation;
(g) urban–rural linkage and national milk grid;
(h) rural upliftment towards self-reliance;
(i) significant contribution to national income foreign exchange revenue and consultancy aspects;
(j) world leader in milk production;
(k) strong industrial base and support;
(l) genetically upgraded indigenous cattle;
(m) presence of strong buffalo base;
(n) development of indigenous milk breeds; and
(o) protection of both producer and consumer interest.

Traditional Milk/Dairy Product

The traditional dairy products in India are broadly five categories :

(a) *Desiccated milk based* products i.e.: Khoa/Mawa, Gulabjamun, Kalajamun Lalmohan, Barfi Kalakdan, Milkcake, Peda, Rabri, Khurchun, Basundi Kulfic, etc.
(b) *Heat Acid coagulated products,* i.e.: Paneer, Chhana, Rasogulla, Rasomalai, Rajbhog, Kheer Mohan, Sandesh, Pantua, Chhana Murki and Chamcham, etc.

(c) *Cultured/Fermented Products* i.e.: Daih, Misti Dahi, Shrikhan, Lassi, Matha/Chacha, Raita, Dahi Vada, etc.
(d) Fet-Rich Products i.e.: Ghee, Ghee reduce, Chocklet Barfi, Makhan, Malai, etc.
(e) *Milk-based Pudding/Desserts,* i.e.: Kheer, Payasam, Phirni, Sevian, Sabodana, Kheer, Lauki Kheër, Sohan Halwa, Gazar Halwa, Kazu Barfi, etc.

In north India with its large milk surplus, *Khoa* is used to make sweets like *Barfi, Gulabjamun, Kalajamun, Kalakand, Peda,* etc., eastern India has a wide range of *Chhana*-based sweet like *Rasogulla* and *Sandesh*. Based India has a range of cultural products based on *Dahi*, such as *Srikhand, Misti Dahi*. South India with limited milk availability has no traditional milk drinking but *dahi* and *buttermilk* widely consumed as part of milk.

Scope for Modernisation

Indian dairy is industry poised for a major breakthrough as a result of application of modern technologies in production of traditional milk products. This development is also having a *'tickled down effect'* on traditional dairy sector which has taken up modernisation of its age old methods and product formulation in following ways:

(i) Inducting appropriate technologies for large scale production.
(ii) Using modern packing and labelling system.
(iii) Evolving a quality assurance system.
(iv) Collecting market intelligence to inspire confidence among them and meet global standards.

Pattern of Milk Utilization

Milk is widely produced in small quantities of 1-4 ltr. by 70 million small and marginal farmers in India in 50,000 villages. Buffalo and Cow and Goat are the main milk animal in India subcontinents. Buffalo contribute 54%, cow 43% and Goat 3% to India's total milk output.

(B) Weakness of Dairy Development in India

The dairy development policy over the years have undergone dramatic changes and suffered from inconsistencies. It consists of following weaknesses :

(a) Inconsistency in livestock development policy and conflicting goals;
(b) Failure to promote buffalo;
(c) Wide mandate and duplication of services;
(d) Presence of large unproductive cattle and buffalo population;
(e) Failure of dairy co-operatives to evoke success in all stages;
(f) Small and scattered herds;
(g) Low per capita availability of milk;
(h) Under-utilised infrastructure;
(i) Unequal distribution of dairy industries;
(j) Variation in demand and supply;
(k) Private milk marketing system;
(l) Competition over dairy products and poor marketing strategy by co-operation;
(m) Lack of professionalism;
(n) Lack of co-ordination and co-operation among the agencies involved in dairy development;
(o) Lack of proper monitoring, controlling and evaluation of the programmes;
(p) Inadequate grassroots level extension system;
(q) Politicisation of dairy co-operation;
(r) Lack of proper surveillance and disease reporting system;
(s) Huge cost involved in animal health care;
(t) Lack of trained technical manpower at the grassroots level;
(u) Social taboos and religious beliefs; and
(v) Inadequate participation of NGOs and R & D Organisations.

(C) Opportunities in Dairy Sector

At present, the dairy development activities are being carried out by national level organisations such as NDDB, ICAR and its Institutes and state level organisations like milk federations, animal husbandry departments and non-government organisations. The success of these organisations in achieving objectives rest with their ability to harness the available opportunities in the best possible way to bring about the desirable change process or improvements in the coming years. These opportunities are :

(i) promoting of buffaloes as milk animals;
(ii) evolving of clear cut national policy;
(iii) eliminating unproductive cattle;
(iv) strengthening of co-operative structures;
(v) milk production by masses;
(vi) less external input and sustainable dairying;
(vii) promotion of farming system approach.
(viii) fodder development and identification of alternative feed sources;
(ix) embryo transfer technology on *in-vitro* fertilisation technique;
(x) tapping of non-conventional fodder resources;
(xi) everything of grassroots level structures;
(xii) tapping of expert market potential;
(xiii) improvement of health care facilities;
(xiv) defining a clear cut breeding policy;
(xv) disease resistant indigenous cattle and buffaloes;
(xvi) draught potential of our indigenous cattle;
(xvii) developing of potential areas/states suitable for dairying;
(xviii) eliminating of disparity among the milk production system;
(xix) protection of weaker section and promotion of dairying among them as a tool for poverty alleviation and self-reliance;
(xx) empowerment of women;
(xxi) utilisation of Indigenous Technical knowledge;
(xxii) promoting co-ordination and co-operation among the development agencies;

(xxiii) self-generation of resources and elimination of foreign aid; and

(xxiv) diversification of research and development (R & D) programme related to dairying.

(D) Threats of Dairy Development

Threats are the situations that hinder the abilities to grow, develop and progress in a given environment. The recent liberalisation policy has paved the ways for the multinational entry in this sector which in turn threatens the indigenous dairy sector. Dairying faces following threats :

(a) competition and entry of multinationals;

(b) high cost concentrates and lack of adequate fodder resources;

(c) huge cattle population and its competition over scarce resources;

(d) scrub bulls and cross-breed males;

(e) disease susceptibility of cross-breeds;

(f) high cost of maintenance of cross-breed cattle;

(g) increased cost of milk production;

(h) urban-rural divide;

(i) existence of unorganised milk marketing system;

(j) lack of remunerative price for milk;

(k) lack of credit availability to farmers and exploitation by money lenders;

(l) exploitation by middleness and traders;

(m) high operational cost of dairy industries and under-utilized plant capacity;

(n) duplicacy of research and uneven development priorities;

(o) non-implementation of stringent quality, control measures in liquid milk marketing;

(p) lack of commitment among dairy development personnel;

(q) poor grassroots level functionaries;

(r) organisational problems; and

(s) prevalence of wide gap and variation in the technology adoption process.

Thus, Indian dairy sector has experienced a sea change during the latter part of the 20th century—it has emerged from developing to a developed dairying nation. By the end of 2005, India has achieved the laurel of being the highest milk producer in the world. The investment, efforts, innovations and energy of our farmers and industry have seen India moving from a state of insignificance to a major dairying nation in the world only due to Operation Flood Programme.

Besides, the tremendous growth in milk production, processing, value addition and marketing, there has been quantum jump in the dairy machinery manufacturing, ancillaries and packaging industries. Till latter part of the last century, India was primarily import dependent. But, due to the remarkable efforts for indigenization by our entrepreneurs in small, medium and large sectors, the country could transform from its import dependent to exporting status.

TABLE 1.1

Value and Volume of Output of the Indian Dairy Industry, 2001

Total Milk Production : 84.6 million tonnes

A. Used as Liquid Milk (46% of total production) : 38.9 MT

B. Used as products (54% of total production) : 45.7 MT

A. Luquid Milk	*Consumption (%)*	*Value of consumption (MT)*	*Rate (Rs./ tonnes)*	*Value (Rs. billion)*
Rural	45%	17.5	9,000	157.5
Urban	55%	21.4	12,000	256.8
Total	100%	38.9	—	414.3

B. Milk Products	*Liquid Milk processed (MT)*	*Volume of Milk products (MT)*	*Rate (Rs./ tonnes)*	*Value (Rs. billion)*
	45.7	11.4	60,000	684.0
Grand Total (A + B)		1,098.3 ($22 billion)		

Source : Dairy India Year Book, 2004, p. 8.

TABLE 1.2
Volume and Value of Market for Traditional Milk Products, 2001

Products	*Volume (MT)*	*Rate (Rs. '000/ tonnes)*	*Value (Rs. (billion)*
Ghee	1.3	100	130
Makkhan	0.4	100	40
Khoa–based sweets	2.0	100	200
Chhana-based sweets	1.0	70	70
Paneer	0.2	90	18
Curd and curd products	6.0	,20	120
Total			578 ($11.5 billion)

Source : *Ibid.*

TABLE 1.3
Value and Volume of Output of the Organised Dairy Sector, 2001

Products	*Volume (MT)*	*Rate (Rs '000/ tonnes)*	*Value (Rs. (billion)*
Liquid Milk	6	12	72
Branded mithais	0.4	100	40
Branded ghee	0.15	110	16
Western products			
Milk Powders	0.35	80	28
Cheese	0.01	150	2
Ice Cream (million litres)	150	50 litre	8
Butter	0.04	120	4
Total			170 ($3.4 billion)

Source : *Ibid.*, p. 8.

TABLE 1.4

Milk Flow–from Producer to Consumer, 2001*

Dairy Animal Population	*Cattle*	*Buffaloes*	*Others*	*Total*
Total (Million)	196	89	Including	285+
@ Milch Animals	54	41	Goats	95+
In–Milk	31	28	etc.	49+
Milk Output (MT)	36.4 (45%)	45.7 (52%)	2.5 (3%)	84.6 (100%)

Production Sector	*Rural*	*Urban*	
Dairy Producers (million)	70	Neg.	70+
Milk Production (MT)	83.0 (98%)	1.6 (2%)	84.6 (100%)

	Rental	*Disposal*
Grass Volume (MT)	30.0 (35.4%)	53.0 (62.6%)

	Pro.	*Non-Pro.*	*Co.-op./ Govt.*	*Pri-vate*		
Volume (MT)	30.0 (35.4%)	9.0 (10.6%)	6.0 (7%)	38.0 (45%)		
Price (Rs./kg.)	7.00	8.00	10.00	10.00	12.00	
Value (Rs. Million)	210,000	72,000	60,000	380,000	19,200	741,200

Grand Total Received By Producers		**Rs. 741.200 Million**	
Consumption Sector			
Population (million)	741 (72.2%)	286 (27.8%)	1,027 (100%)
Per Capita Milk Availability	144 (Rural)	437 (Urban)	
Gross Volume (MT)	39.0 (46%)	45.6 (54%)	84.6 (100%)

			Organised		Traditional	
			Co.op/ Govt.	Pri- vate		
Volume (MT)	30.0 (35.4%)	9.0 (10.6%)	6.0 (7%)	4.6 (55%)	35.0 (41.5%)	84.6 (100%)
Liquid Milk	#30.0	#9.0	5.0 (83%)	1.0 (22%)	23.0 (66%)	$68.0 (80.4%)
Milk Products (Milk eqvt.)			1.0 (17%)	3.6 (78%)	12.0 (34%)	16.6 (19.6%)
Ghee			0.03	0.10	1.17	1.3
Khoa/Chhena/Paneer			0.10	0.10	1.80	2.0
Milk Powder			0.10	0.25		0.35
Table Butter			0.03	0.01		0.04
^Other Products			0.60	0.40	0.50	1.5
Price (Rs./kg.)						
Liquid Milk	7.00	8.00	12.00	12.00	12.00	
Ghee			120.0	120.0	110.0	
Khoa/Chhena/Paneer			70.00	70.00	70.00	
Milk Powder			80.00	90.00		
Table Butter			120.0	120.0		
^Other Products			100.0	100.0	100.0	
Value (Rs. million)						
Liquid Milk	210,000	72,000	60,000	12,000	276,000	630,000
Ghee			3,600	12,000	128,700	144,300
Khoa/Chhena/Paneer			7,000	7,000	126,000	140,000
Milk Powder			8,000	22,500		30,500
Table Butter			3,600	1,200		4,800
^Other Products			60,000	40,000	50,000	150,000
Total (Rs. million)	210,000	72,000	142,200	94,700	94,700	580,700
Grand Total paid by consumers	Rs. 1,099,600 million					

Neg.=Negligible, @ Includes, dry, but excludes those of calved even once.

Includes conversion into products by housewives, halwais, etc.

$. Includes value for milk consumed by rural products,

^ Ice creams, processed cheese, etc.; * Estimated.

Source : *Ibid.*

TABLE 1.5
Milk Production and Availability

Year	*Milk Production (MT)*	*Per Capita Availability (gm per day)*
1960	20.0	124
1970	22.2	114
1980	31.6	128
1990	53.9	178
1995	66.2	197
1996	69.1	202
1997	71.9	207
1998	75.2	213
1999	78.1	214
2000	01.0	219
2001	84.6	226
2005	98.3	243
2010 (Projected)	119.6	274

Source : Dairy India, Yearbook.

The growth projections for their demand in the organised sector are presented in Table 1.6.

TABLE 1.6
Projected Demand for Major Milk Products in the Organized Sector, 1988–2009

(*Metric Tonnes*)

Product	*Demand 1988*	*Projected demand 2009*
Ghee	100,000	200,000
Gheese	4,200	15,000
Paneer	1,000	16,000
Shrikhand	3,000	5,650
Rasogolla	1,600	6,000
Gulabjamun	3,000	5,850

The National Sample Survey (NSS 50th round) data support the rising trend in the average monthly per capita expenditure on milk and milk products as percentage of the expenditure on food (Table 1.7). Interestingly, such expenditure in rural households in Northern India is generally higher (20.3–42.5 percent) than urban households.

TABLE 1.7
Share of Milk and Milk Products in Average Monthly Per Capita Food Expenditure

(*Percentage*)

Region	*Average monthly per capita expenditure*
Northern States	22.8 – 34.1
Western States	17.4 – 23.2
Southern States	10.4 – 14.7
Eastern States	8.5 – 13.9

Source : NSS Consumer Expenditure Survey, 50th Round (July 1993 to June, 1994.

A recent study by the National Dairy Research Institute (NDRI) has revealed the varying share of per capita expenditure on milk and milk products in the total expenditure on food by urban, semi-urban and rural people in northern and southern regions, as shown in Table 1.8.

TABLE 1.8
Percentage Share of Milk and Milk Products in Per Capita Food Expenditure

Region	*Type of Population*		
	Rural	*Semi-Urban*	*Urban*
Northern	39.5	27.2	24.7
Southern	13.7	19.3	16.0

The actual sample for the study consisted of 250 retailers in four metro cities. In addition, a similar survey was conducted in

Kolkata. The city-wise break-up of these retailers is presented in Table 1.9.

TABLE 1.9
Break–up of Retailers Surveyed in Five Cities

Metro/Cities	*Number of Sample outlets*
Mumbai	89
Delhi	95
Bangalore	40
Ahmedabad	26
Sub-total	250
Kolkata*	400
Total	650

* Conducted in different framework, guided by. the author.

Market size of traditional milk sweets for five cities was projected, based on average sale per retailer multiplied by number of retailers in the metro/city in following Table 1.10.

TABLE 1.10
Estimated Market Size of Traditional Milk Sweets in Selected Cities

City	*Market size*	
	Value (Rs. million/ annum)	*Quantity (Tonnes/ annum)*
Ahmedabad	1,701	17,799.7
Bangalore	2,571	24,169.6
Delhi	23,666	254,604.5
Mumbai	10,777	88,726.1
Kolkata*	15,862	—
Total	54,577	

* Market estimates available in value only.

The non-organized sector dominates the market with a share of around 93.6 percent in Table 1.11.

TABLE 1.11

Market Share of Non-organized and Organized Sectors in Kolkata

(Percentage)

Product base/process	*Market share (based on value)*	
	Non-Organized Sector	*Organized Sector*
Khoa–based	91.7	8.3
Chhena–based	96.3	3.7
Cultured products	93.0	7.0
Other products	67.0	33.0
Overall	93.6	6.4

India as blessed with several competitive advantages, one of which is the highly skilled, unskilled and expensive manpower resource to cater to the ever increasing manpower demands, both at the national and global levels.

In the 21st century, an important watershed of our industrial evolution is expected to take place, wherein agro-industry and especially dairying would play a pivotal role with the increasing pull of the market *vis-a-vis* push of production, significants, opportunities are knocking at our doors, for both domestic and export market.

INDIA AS THE LARGEST DAIRY MARKET OF THE WORLD

India today is the world's largest and fastest growing market for milk and milk products. With an annual growth rate of about 4.5 percent, the country's milk production, mostly rural-based, exceeded 230 million litres per day (84.6 million tonnes per year) in 2001.

For sheer number, Indian dairying has no match in the world. The figures are simply mind boggling. For example : Some 70 million farmers maintaining a milch herd of about 100 million—54 million cows and 42 million buffaloes, fed largely on crop residues. They account for 97 per cent of all milk produced starts as a trickle of one of two litres per family in some 500,000 remote villages. A unique collection system transforms this trickle into a veritable flood of 120 million litres per day for urban consumers.

India's White Revolution owes much to the Anand Pattern of cooperative dairying. Its network covers over 10 million farmer-members in 96,000 village societies in 170 milksheds spread over 270 districts of the country. *Operation Flood* (1970–96) has modernized India's dairy sector. In 2002, over 30 dairy plants have been awarded the ISO : 9000 and HACCP certification and their number is increasing.

Marketing

The indigenous dairy products are India's largest selling and most profitable segment after liquid milk and account for 50 percent of milk utilization. Significant headway has been made in the industrial production of traditional sweets such as *shrikhand, gulabjamun, peda and burfi*. India's dairy market is multi-layered, shaped like a pyramid with the base made up of the vast market for low cost, liquid raw milk. The narrow tip at the top is a small but affluent market, largely for western-type and fresh packaged dairy products.

The bulk of the demand for milk, however, is in urban areas, amounting to some 125 million lpd, accounting for more than 80 percent of traded milk. Herein lies the immense growth potential of the modern organized sector. Presently, the modern milk distribution network supplies hygienically-packed, quality pasteurized milk to about 1,000 cities and towns. This number could go up by almost five times in the foreseeable future. According to one estimate, the packed, pasteurized, liquid milk segment, presently estimated at 15 million lpd, would double in the next five years, giving both strength and volume to the modern sector.

The effective milk market is largely confined to urban areas, inhabited by 285 million people. Over half of India's total milk production is consumed in urban India. The urban population is projected to cross the 400 million mark by 2011. The expected rise in the purchasing power of growing urban population would give an added boost to the dairy market.

New emerging dairy markets will focus on :

(a) Food Service Institutional Market : It is growing at double the rate of consumer market;
(b) Defence Market : An important growing market for quality products at reasonable prices;
(c) Ingredients Market : A boom is forecast in the market of dairy products used as raw material in pharmaceutical and allied industries; and
(d) Parlour Market : The increasing 'away-from-home' consumption trend opens new vistas for ready-to-serve dairy products which would ride piggyback on the fast food revolution sweeping urban India.

In the new millennium, the Indian dairy industry is gearing itself to face new challenges and opportunities. Its strength lies in the synergistic partnership forged between the farmer and the dairy professional. This has already given a thrust to the dairy industry to emerge as a full-fledged agri-business.

TABLE 1.12
Projected Population and Milk Availability, 1999-2010

Year	*Projected population (million)*	*Annual Milk production at 4% growth rate (million tonnes)*	*Per Capita Availability*	
			Kg./year	*gm./day*
1999-2000	1,002.1	77.69	78.23	214.32
2000-01	1,017.5	80.80	79.87	218.82
2001-02	1,033.0	84.03	81.55	223.42
2002-03	1,049.1	87.39	83.26	228.11
2003-04	1,065.8	90.88	85.01	232.90
2004-05	1,092.8	94.52	86.80	237.81
2005-06	1,100.0	98.30	88.62	242.79
2006-07	1,117.3	102.23	90.49	247.91
2007-08	1,134.3	106.32	92.38	253.09
2008-09	1,151.2	110.57	94.32	258.41
2009-10	1,168.1	115.00	96.31	263.86

Source : Expert Committee on Population Projections, Planning Commission, Department of Animal Husbandry and Dairying, Ministry of Agriculture, Government of India.

Table 1.13
Trends in India's Annual Milk Production and its Per Capita Availability, 1960-2010

Year	*Milk Production (MT)*	*Per Capita Availability*	
		Kg/year	*gm/day*
1960	20.0	45	124
1970	22.2	42	114
1980	31.6	47	128
1990	53.9	65	178
1995	66.2	72	197
1996	69.1	74	202
1997	71.9	76	207
1998	75.2	78	213
1999	78.1	78	214
2000	81.0	80	219
2001	84.6	82	226
2005	98.3	89	243
2010	119.6	100	275

Source : Department of Animal Husbandry & Dairying, Ministry of Agriculture, Government of India.

Table 1.14
Production of Milk by Species in India and the World, 2001

Species	*India*		*World*	
	Production (MT)	*Percentage*	*Production (MT)*	*Percentage*
Buffalo	45.7	54	69.2	11.8
Cow	36.4	43	493.8	84.5
Goat etc.	2.5	3	21.6	3.7
Total	84.6	100	584.6	100

Source : Department of Animal Husbandry and Dairying, Ministry of Agriculture, Government of India and FAO, Rome.

TABLE 1.15
Value of Output from Livestock and Crops, 1993-99

	At current prices				*At 1993–94 prices*			
	1993-94	*1995-96*	*1997-98*	*1999-2000*	*1993-94*	*1995-96*	*1997-98*	*1999-2000*
I. Livestock	669	858	1071	1302	669	720	766	831
Milk group	434	572	707	904	434	475	507	559
Meat group	125	154	198	209	125	128	134	140
Eggs	23	28	34	40	23	25	26	29
Dung	62	74	90	101	62	64	67	68
Other*	25	30	42	48	25	28	32	35
II. Crops	2049	2567	3196	4056	2049	2122	2257	2463
III. Value of output from Agriculture (I + II)	2718	3425	4267	5358	2718	2842	3023	3294
GDP	7813	10733	13900	17865	7813	8996	10163	11520

*Includes silkworm cocoons, honey, wool and hair.

Source : National Accounts Statistics, Central Statistical Organisation, Ministry of Statistics and Programme Implementation, Government of India.

TABLE 1.16
Value of Output Milk and Livestock as Percentage of GDP, Agriculture and Crops, 1999–2000

(*Percent*)

	Base Year 1993–94	*1999–2000 Fixed Prices*	*1999–2000 Current Prices*
Milk group output as percent of			
Livestock	64.8	67.2	44.1
Crops	21.1	17.3	22.2
Agriculture	14.8	16.3	9.8
GDP	4.2	3.8	2.4
Livestock output as percentage of			
Crops	32.6	25.8	31.9
Agriculture	22.9	24.2	22.2
GDP	6.5	5.6	5.5
Agriculture as percentage of			
GDP	28.4	23.1	24.9

Source : National Accounts Statistics, Central Statistical Organization, Ministry of Statistics and Programme Implementation, Government of India.

TABLE 1.17
Value of Output of Milk and Livestock : 1993-99

(*Rs. Billion*)

Year	*1993-94*	*1996-97*	*1999-2000*
Milk Group			
at current prices	434.1	643.3	903.6
at 1993–94 prices	434.1	494.5	558.9
Livestock			
at current prices	669.7	971.6	1302.3
at 1993–94 prices	669.7	745.6	830.8

Source : National Accounts Statistics, Central Statistical Organization, Ministry of Statistics & Programme Implementation, Government of India.

Among the job generating potentials of the different sectors, the dairy sector is always reckoned as priority sector. Dairying is not only the largest job-generating sector but it also provides round the year employment-*cum*-safety nets to the weaker sections of our society.

Significant opportunities are now emerging for enhancing organised milk production, processing and marketing. The government policies, over the years, are becoming congenial for investment in the dairy sector according to Dr. Animesh Banerjee. The President of India Dairy Association 2004 (*India Dairyman*, 56, 10, 2004, p. 21).

But according to Sri Bhairon Singh Shekhawat, Hon'ble former Vice-President of India, "We are now living in the era of globalization. The new globalization economic world order, no doubt, offers, opportunities for increased export led growth. However, there is a need to protect and safeguard against the unhealthy competition from rich and highly subsidized farmers of the developed world. The best safeguard of course lies in building intrinsic competitive strength in our own industry. We should ensure the expansion of infrastructure facilities. We need the big quantitative improvement, right from the primary production point to the marketing end with our outputs conforming to the best of our international standards of quality and reliability" (*Indian Dairyman*, 56, 10, 2004, p. 12).

In the words of Dr. (Mrs.) Annita Patel, Chairman, National Dairy Development Board, "Today we see enormous changes in the competitive market environment of players engaged in the business of processing of milk has expanded manifold and this has thrown up both challenges and opportunities for all the responsible players, we should be producing more than 130 million tonnes of milk by 2015.

It is desirable that as 'the milk' production grows the share of the organized sector increases. At the same time, we must recognize that processing through modern dairy plants results in milk and milk product being supplied at the higher cost than the traditional system because of the better practices required to be followed in procurement, processing and marketing which will depend upon increasing consumer's purchasing power and how effective we are in educating consumers on the benefits of hygienic high quality milk. This in turn will be possible only when government is able to enforce the existing food laws to ensure quality and hygiene.

The essence of the challenge we face is to make certain that our dairy industry continue to keep its purpose constant, that dairying continues to contribute significantly to the income and quality of the lives of the tens of millions of our fellow citizens who depends on it. If we are to meet the challenges, we can only do so as an industry. At the same time, there is a genuine need to modernize our industry. And thus, requires linking producers directly with market through professionally managed institutions.

Co-operatives which continue to be the dominant player, have thus, a special responsibility. They can't afford to lose any time in removing the shackles of the past where necessary adopting a professional approach on that is the only route to survival.

Costs need to be reduced both through efficiencies in operations and economies of scale to ensure that co-operative not personal or political gain is the only way to success.

Milk Market

C.H. Eckler, Combs and Herold Macro have made a detailed analysis of milk markets. Over 35% of the total milk produced is utilized as fluid milk or cream. Pasteurization, mechanical refrigeration and the use of methods based upon the understanding of sanitary and scientific questions involved have made this possible.

Butter

About 20 % of the milk is used for making butter. During the second world war butter declined in relative importance because under the existing price controls and rationing the demand for other dairy products drew the available milk supplied away from butter. In 1949, 141 crore (pound) of creaming butter were manufactured in United States, another 30 crore $ (pound) of butter were made on farms in the same year. During recent years the amount of farm-made-butter has declined rather rapidly. The principal butter producing area of the United States in the West North Central section including Minnesote, Lows, Missouri, North and South Dakota, Nebraska and Kansas.

Cheese

The amount of milk used for cheese is relatively small in comparison to butter making. In 1949 its percentage was only 9.3% of the total milk produced. The average cheese factory in United States is small as compared with other factories.

Concentrated Milk

Gail Borden was awarded a patent on process of producing *condensed milk* is 1856 who established first condensed milk factory in wolecottville, Litchfield country, Conn, in same year. Concentrated milk is prepared for market in various farms. This dried milk has become an important factor in preservation of the solid constituents of large quantities of whole milk, skim milk, butter milk and whey for human food or feed for animals. In 1948 total dry milk production was 100 crore $ (pound). Dry skin milk represented over 65% of total milk.

Ice-cream

The Ice-cream industry produces more than 50,000 gallons of all types of frozen dairy products annually. The chief products of the ice-cream factory are ice-cream, *sherbet* and ices. The raw products required for ice-cream.

If we strive to seek excellence we need to urgently re-examine our existing institutional structures and build, into these the capacity to make constant changes as the environment demands and a leadership of a high order that provides the dynamism and flexibility to catalyse the progressive change that we seek.

In this liberalized environment a modern dairy industry can only thrive if it competes effectively. But the players need to co-operate to be able to build a secure foundation for the future of the industry. A critical area in which they need to co-operate is sharing information on the procurement, processing and marketing of milk and milk products.

Deregulation, privatization, changes in consumer preference in regard to price and global quality standards, information boom and growing demand for transparency and accountability are the main challenges being faced by Indian industries. Also the native of work, workforce, workplace and psychological environment in organisation in undergoing change bringing both opportunities and challenges, one of the challenges is to excel and create a position in the market to sustain the growth and development for survival.

Competition has also necessitated improving the quality of products, reduction in cost, innovation in product and development and enhancement of productivity leading to achieving corporate excellence. In this emerging global scenario, the focus should be on achieving management excellence through enhancing access to information technology, modernization and technology upgradation for raising productivity, reducing costs, improving quality standards targeting at retaining competitive edge both in the domestic and export market and trained, motivated, goal directed and committed work force. The new environment demands creative management, redesigning and remodelling of traditional system and practices to suit the new circumstances (Juneja, 1999).

According to the former President of India Dr. Kalam, "the leader should acquire upto date knowledge through continuous education and skill training."

To nurture the personal and organizational excellence in Indian Dairy industry, we need to first understand the scenario. Indian Dairy industry is characterised by the following constrains:

(i) The productivity of our milch cattle is very low (1.5 litre/animal/day).

(ii) The advantages of buffalo have not been fully realised.

(iii) Quantitative and qualitative shortage of fodder resources—as much as 10-15% increases can be recorded in the existing milk production through adulated feeding of bovine production.

(iv) There is under non-rishment of heifer calves, which contributes to calf mortality, infertility and delayed conception.

(v) Only 10% of the milk produced is in the organised sector, leaving the remaining portion in the hands of milkmen, providing chances for adulteration and exploitation.

(vi) The marketing infrastructure relating to transportation, processing, packaging and distribution of milk from production centre (rural area) to consumer centres (urban-areas) are ill-equipped.

(vii) Pesticide/insecticide residues, adulteration with synthetic products, low keeping quality, bacterial contamination and lack of brand promotion will be detrimental for Indian dairy products marketed across the seas.

(viii) The latest technologies such as use of BST (Bovine Samatotrofin), E.T. (Embryo Transfer), etc. have not been applied to its full potential. As a result, it remain an academic exercise in most part of our country.

(ix) Dairy being a part fragmented industry required high capital investment to break his barrier of being fragmented either in processing technology or in the specialised logistics and ware housing (*Kurukshetra,* Vol. 54, No. 2, p. 34, 2005).

The performance of dairy sector in India and particularly in Bihar over the past three to four decades has been extremely impressive. The future of this sector is promising as there is a huge domestic demands as well as better scope for the export dairy products. Dairy opens an array of business opportunities and marketing. The recent liberalisation has thrown a bagful of opportunities for dairy entrepreneurs.

2

White Revolution : Needs, Objectives and Significance

Milk is the principal source of animal protein for majority of the people in this century. A study of *white revolution* is of great value. Proteins of high biological value are derived mainly from milk which are essential for infants, nursing mothers, old men, patients and manual workers. The value of milk and milk products in the human diet is universally recognized. Milk has occupied a prominent place in the dietary of Indian people from time immemorial. Dairy is a significant source of generating rural income and employment. It provides much more stable and continuous economic base than crop production and security against the vicissitudes of drought and famine.[1]

White Revolution refers to a revolution of milk production in India and also in Bihar. The term White Revolution is being used for the planned development of dairy farming and dairy industry.[2]

Milk and its products are traditional and highly acceptable form of animal products in India, the staggering pace of growth of white revolution is based on veterinary education. As early as *vedic period*, veterinary science was developed in India. *Atharveda* and *Mahabharat* (1500–500 BC) have references as dairy farming, study of animal anatomy, horse management and health care aspects. The existence of veterinary hospitals have been well documented in Sanskrit script and edicts in different periods from the first milk millennium.[3]

'White Revolution' is quite relevant in Indian economy particularly at the time of globalisation. It has travelled a long way from yester years when the dairying and milk supply schemes in the states like Bihar, U.P., M.P., Gujarat, etc. were implemented by the milk Commissioner, the Director of Animal Husbandry and the Registrar of Co-operative Societies. Dairying of late in the state like Bihar has assumed new dimensions as a strategy for rural development for eradicating poverty and under employment of the small and marginal farmers and agricultural labour.

Since from the chronic shortage of milk, India has emerged today as the largest producer of milk in the world. This has been achieved through *Operation Flood*/White Revolution, one of the worlds's largest dairy development programmes which has created strong linkages between the rural producers and urban consumers. All this happened under *autarky* and regulated domestic markets. Since 1990s, India embarked upon liberal policy framework, which got reinforced with the signing of Uruguay Round Agreement (URA) in 1994.[4] This increasingly exposed the Indian dairy sector to the world markets, which have been distorted to by policies of high tariffs, domestic support and export subsidies in the developed countries.

ISSUES OF WHITE REVOLUTION

This study revolves around the following issues :

(i) Where does Indian dairy industry stand in terms of competitiveness/efficiency and what are important factors affecting competitiveness of the industry ?

(ii) What are the main provisions of the WTO Agreements relevant for the dairy sector and whether WTO has been successful in reducing trade barriers and market distorting subsidies and therefore meeting the objective of attaining greater benefits from trade liberalization?

(iii) What are the options for India in the coming round of multilateral trade negotiations, give the scores of distortions that plague the world dairy market?

No doubt, there is a greater need of the review of White Revolution particularly since the initiation of policy of economic reform and globalisation. As we know in the early 1990's the Government of India initiated major trade policy reform, which favoured increasing privatization and liberalisation of all sectors of the economy and dairy sector was no exception to this.

Dairy sector, particularly, the handling, processing and marketing of fluid milk, which was reserved mainly for two co-operative sectors, was delicenced in June 1991. The private sector companies including multinational were allowed to set-up milk processing and product manufacturing plants. The basic philosophy underlying delicensing was to encourage the competition in procurement and marketing of milk, thus, increasing value for both producers and consumers. Although delicencing attracted a large number of players, concerns on issues like excess capacity, sale of contaminated/substandard quality of milk, etc. induced the Government to promulgate the milk and milk products order in 1992, some provisions of which were again modified in April 1993. Under White Revolution milk and milk products order regulates milk and milk products in the country.

The second related reason underlying this study stems from India's signing of the Uruguay Round Agreement (URA) of the General Agreement on Tariffs and Trade (GATT) in 1994 and becoming member of the World Trade Organisation (WTO) in 1995. Both these developments indicate that sooner or later, the Indian dairy sector will have to face the world dairy markets.[5]

Prior to delicensing of the dairy industry, the trade in dairy products were restricted through various measures (canalization, licensing, quotas, etc.) and the Government of India has designated NDDB as a canalizing agency for all dairy imports who have also expected to co-ordinate export of milk products. However, after the opening up of the industry and as part of commitment under WTO all quantitative restrictions and other non-tariff barriers on the import and export of dairy products were removed and most dairy products were put under Open General License (OGL). From April 1, 2001 all dairy products have been shifted to the OGL with low tariff rates compared to very high tariff rates in developed countries. Subsequent to de-canalization, exports of dairy products are freely allowed provided these units comply with the compulsory inspection requirements of concerned agencies like *Export Inspection Council* (EIC), *Bureau of Indian Standards* (BIS), etc. The above developments exposed the Indian dairy industry to the world market which could have significant implications for milk production and processing sector in India.

World trade in dairy products has been distorted to for decades by both domestic and international trade policies. Inspite of regional and multilateral trade agreements dairy policies continue to distort resource use and world market prices. The level of producer subsidy equivalent has remained high in most of the developed countries throughout the 1980s, and 1990s. Most of the developed countries provide heavy export subsidies under the export subsidy programmes under green box, blue box and amber box, to increase their exports, which create unhealthy competition in the domestic markets.[6]

The earlier attempts at White Revolution dairy development can be traced back to the British Rule, when the defence department established military dairy farm to ensure supply of milk and milk products to the colonial army. The first farm was established in Allahabad in 1913.[7] (A. Banerjee, 1999, *Export Potential of Indian Dairy Products,* Indian Dairyman, pp. 525-32).

In the post-Independence period modernization of Indian dairy industry became a priority of the government with the initiation of a planning process. One of the first policy initiatives immediately after Independence was the recommendation from Milk Sub-Committee on the policy committee on Agriculture, 1950 which resulted in a city Milk Scheme set-up in Delhi, which grew eventually to cover nearly 100 towns and cites by 1960[8] (Parthasarthy, 2001).

These were essentially demand driven consumer-oriented initiatives run by governments and could not compete with milk vendors, as there was no supporting strategy to cover milk producers in rural areas with remunerative prices and other required incentives in an integrated manner.[9]

The State Governments and Central Government formally put through a National Breeding Policy in 1962. The Breeding Policy covered :

(a) selective breeding of the pure Indian cattle breeds,
(b) selective breeding of the pure Indian draught animals,
(c) selective breeding of dual purpose breeds for improving both milk and work output, and
(d) grading up of non-discript Indian breeds with exotic breeds.

MAIN OBJECTIVES OF WHITE REVOLUTION

The main objectives of the White Revolution are[10] :

(i) to promote dairy sector uniformly and effectively,
(ii) to promote dairy science and practice and diffusion of technical know how,
(iii) to design and to plan any dairy or associated plant,
(iv) to advice dairy management,
(v) to promote quality control of milk and milk products,
(vi) to organise technical programmes for training of products,
(vii) to preserve and improve cattle for the greater production of milk and/or increase of draught power,
(viii) to provide consultant's service to any dairy,
(ix) to advice an price fixation, price policy, public relations and allied subjects relating to dairies,
(x) to charge fees for services rendered,
(xi) to establish a provident fund for the benefit of the employees of the Board,
(xii) to sell, mortgage or exchange and otherwise transfer all or any portion to the properties of the Board,
(xiii) to employ such staff as may be necessary for the proper performance of any or all of these functions, and
(xiv) to adopt and undertake any other measure or perform any other duty as may be required by the Government of India by a State Government or which Board may consider necessary or advisable in order to carry out any of the objects of the Board,

According to G.S. Rajorhia, an eminent Dairy Technologist, the needs and objectives of the White Revolution are mainly concentrated in the areas of :[11]

(a) Khoa and Khoa-based sweets (Burfi, Peda, Kalakand, Gulabjamun).
(b) Chhena and Chhena-based sweets (Sandesh, Rasogulla and other Bengali products).
(c) Industrialisation of Ghee making.
(d) Fermented milk products like Chakka, Srikhand, long life Dahi and Misti Dahi.
(e) Formulated/Convenience indigenous dairy products in dry form, such as Gulabjamun, mix powder, Burfi mix, Khoa powder, Kulfi mixed powder, Rasgulla mix powder, carrot milk food etc., and
(f) Manufacture of it can sterilised Kheer, Rabri, Basundi, etc.

Since India has emerged as the World's largest milk producing country because India's milk production during 2002-03 has risen to an estimated annual figure of 91 million tonnes from 81 million tonnes, 79 million tonnes and 77 million tonnes of the previous three years respectively due to white revolution. It is projected that during 2009-10 the milk production in our country would reach 115 million tonnes at an expected growth rate of 4% in milk production as obvious from Table 2.1.

Moreover, Indian agriculture is an economic sybosis of crop and cattle production. The activists of white revolution are predominantly performed by landless labourers and marginal farmers with small holding of one or two milch animals. Such small holders have about 70% of the total milch animals. Nearly 70 million farm families in the country depend on White Revolution for income generation and productive employment for family labour.

Dependence on dairying/White Revolution for the generation of income and productive employment is increasing in the sector of small marginal farmers, while the number of small holdings are on the rise.

Successful implementation of white revolution especially Operation Flood Programme has created a vast countrywide network of co-operative systems. They have played a major role

TABLE 2.1
Annual Milk Production at the 4% of Growth Rate

Year	*Annual Production (in million tonnes)*
1999-00	77.69
2000-01	80.80
2001-02	84.03
2002-03	87.39
2003-04	90.88
2004-05	94.52
2005-06	98.30
2006-07	102.23
2007-08	106.32
2008-09	110.57
2009-10	115.00

Source : *Indian Dairyman*, Vol. 54, No. 2, 2002, p. 99.
Economic Survey, 2005-06

in transmitting urban market consumer demand in rural producers, thus enabling then to receive a more sustainable price for their milk produce as compared to any other agricultural product. This has provoked an increase in the number of farmers taking up dairying as an occupation, thereby contributing to increased milk production.

Jain (*et. al*. 1990), rightly estimated that there will be 69 million breedable cows and 43 million breedable buffaloes in our country by 2000. Of these, barely 10% cows are crossbred. Yet, their contribution of the total cow milk production is about 22%. The crossbreeding native cattle with exotic breeds for the enhancement of milk production gained momentum during the last three decades in some parts of the country.[12] According to FAO, the production growth in India is increasing through improved yield per animal rather than growth in animal number. The global average milk production per cow has been static since 1990. In India, however, it has increased from 520 kg. per year in 1990 to 580 kg. per year. It was anticipated that the projection in the year 2000 would be 775 kg. per year (*Dairy India*, 1977).[13]

A marginal increase in productivity per animal is bound to create radical increase in milk production as India has the largest number of milk animals in the world. The overall growth rat of milk is estimated to be 1.52% per annum (1995-2005). On the country, India had a higher growth rate of 5.41% per annum. It is expected to fall to some extent. Taking 1995 as the base year, it we expected to increased at 3.97% by 2005[14] (*Source* : Indian Dairyman, 53, 10, 2001, p. 34). According to Economic Survey 2005–06 (p. 159) the milk production and per capita availability of milk have rapidly increased between 1950–51 to 2004–05. The milk production increased for 17 million tonnes in 1950–51 to 90 million tonnes in 2004–05 whereas per capita availability of milk also increased from 124 gms. per day in 1950–51 to 232 grams per day in 2004–05.

So far as percapita availability of milk for consumption is concerned, with the growth of human population in India at a compounded rate of 1.86% and the annual growth rate in milk production at 4%, the country will attain self-sufficiency by 2007-08 in providing 250 gm. of milk to its population. Details of projected production and per capita availability of milk are presented in Table 2.2.

Table 2.2
Per capita Availability of Milk

Year	*Annual Milk production at 4% growth rate in (million tonnes)*	*Percapita Availability of milk (kg/year)*	*Per capita Availability of of milk per day (in gm.)*
1999-00	77.69	78.23	214.32
2000-01	80.80	79.87	218.82
2001-02	84.03	81.55	223.42
2002-03	87.39	83.26	228.11
2003-04	90.88	85.01	232.90
2004-05	94.52	96.80	237.81
2005-06	98.30	88.62	242.79
2006-07	102.23	90.49	247.91
2007-08	106.32	92.38	253.09
2008-09	110.57	94.32	258.41
2009-10	115.00	96.31	263.86

Source : *Indian Dairyman*, Vol. 53, 10, 2001, p. 34.

National Sample Survey (NSS) data on the consumption of milk and milk production shows short of domestic milk production as it does not include a large part of institutional demand for milk.

The National Dairy Development Board (NDDB) has also projected that milk production will reach 3,330 lakh kg. per day by 2009-10 from 1950 lakh kg. per day in 1997-98 (NDDB Vision 2010). It has also estimated that production would increase to 2665 lakh kg. per day by 2009-10 from 1560 lakh kg. per day during 1997-98 in the *Operation Flood* project area. Considering the retention of 45% the projected marketable surplus may rise from 860 lakh kg. per day to 1480 lakh per day from 1997-98 to 2009-10 as indicated in Table 2.3.

TABLE 2.3
Projected Milk Production Surplus of Milk in India

Year/Particulars	*1997-98*	*2003-04*	*2009-10*
1. Total production (lakh kg. per day)	1950	2525	3330
2. Production of Projected Area (–) lakh kg. per day @ 80% of Production	1560	2025	2665
3. Retention	45%	45%	45%
4. Marketable Surplus (–) (Lakh kg. per day)	860	1120	1480
5. Procurement (–) (Lakh kg./day	130	230	535
6. Percentage of Procurement to Marketable Surplus	15%	21%	36%

Source : National Dairy Development Board Vision, 2010, *Indian Dairyman*, 54, 2, 2002, p. 100

The level of income is the most important determinant of the demand for milk and milk products in the field of white revolution. It is to be noted that the price of milk and milk products also influences the demand for them. However, prices of other related products may not significantly influence the demand for milk and milk products. The income elasticity of demand for milk and milk products has reduced briskly over the years with an increase in per capita national income. It is likely to stagnate during the coming decade.

The expenditure on an milk and milk products varies from region to region. People living in the northern parts of the western region spend a higher proportion of their income on milk and milk production as compared to people from other parts of the country as presented in Table 2.4.

TABLE 2.4

Per capita Demand for Availability of Milk in Kg/year at Per capita Growth Rate of National Income @ 4%

Year	*Projected Population (Crore) (Rs.)*	*Per capita National Income*	*Per capita Demand for Milk (kg./year)*	*Per capita Availability of milk (kg./yr.)*
1999-00	100.21	3,111.00	76.9	78.23
2000-01	107.75	3,236.00	77.5	79.87
2001-02	103.30	3,365.00	78.0	81.55
2002-03	104.91	3,500.00	78.2	83.26
2003-04	106.58	3,440.00	78.3	85.01
2004-05	108.28	3,786.00	78.3	86.80
2005-06	110.00	3,937.00	78.1	88.62
2006-07	111.73	4,094.00	77.7	90.49
2007-08	113.43	4,258.00	77.2	92.38
2008-09	115.12	4,429.00	76.5	94.32
2009-10	116.81	4,606.00	75.7	96.31

Source : Expert Committee on Population Projection, Planning Commission (as shows in Government of India, 1999, *Indian Dairyman*, 54, 2, 2002.

The increasing disposable income of this strong middle class will contribute to the speedy growth of market for milk and milk production. The demographic structure of market is expected to change dramatically in the next five years. Rural India too, will emerge as a potential market for many packaged milk products. It is to be noted that the income elasticity of demand for milk in rural markets is also relatively higher as compared to urban areas.

CONSUMPTION PATTERN

The Western and Northern regions of the country consume a major amount of milk is direct or indirect liquid form. On the other hand, milk consumption in the southern and eastern regions is largely through the tea and coffee. Out of the total production of milk 46% is consumed as fluid milk and 7% of curds. Details of the different forms of milk consumption are presented in Table 2.5.

TABLE 2.5
Consumption in Terms of Percentage of Different Form of Milk

Products	*Consumption in (%)*
1. Liquid Milk	4.60%
2. Ghee	28.00%
3. Curd	7.00%
4. Butter	6.58%
5. Khoa	5.50%
6. Milk Powder	3.8%
7. Cheese	2.0%
8. Others	1.12%

Source : Indian Statistical Institute, 1997.

Milk dominates the consumption of livestock products across the income levels. It is over four times more important than any other livestock product. The per capita consumption of milk at the annual growth rate of about 3.5% would reach 225 ml/day by 2010 (*Dairy India* 1997, p. 10). The market size of fluid milk and ice-cream are expected to grow at the relatively higher rate of 8-10% and 10-15% respectively, as compared to other major dairy products as shows in Table 2.6.

TABLE 2.6
Annual Growth Rate of Milk Products in %

Products	*Annual Growth Rate (%)*
1. Liquid Milk (including flavoured milk)	8-10%
2. Banded Traditional Sweets	5%
3. Ghee	5%
4. Baby Food	3%
5. Skin Milk Powder	6%
6. Ice-cream	10-15%
7. Butter	8-10%
8. Navy Whites	10-15%
9. Whole Milk Powder	2-3%
10. Cheese	10-15%

Source : *Dairy India*, 1997, p. 19 and *India Diaryman*, 54, 2, 2002.

LIQUID MILK MARKETING

Indian economy has now moved from an era of milk scarcity to an era of abundance due to white revolution. Fluid milk offers an exciting marketing opportunity compared to any other milk product due to its developing demand. Selling milk as milk itself is more profitable than convering it as products. Developing new products involves extremely high investment on marketing and the rate of success in the market is low. Added to that is the risk factor.

About 35% of milk product is consumed by urban areas out of this the approximate share from co-operative's supply is 20%, private 2% and government managed dairies 2%. About 15-20 years ago milk marketing was confined to physical distribution. Owing to scarcity of milk marketing it was confined to home delivery.

NEEDS OF WHITE REVOLUTION

White Revolution/Dairy development consists of following needs :

1. Milk producers have to be supplied an improved variety of feeder's seeds and feeder at a reasonable process.
2. Milk selling at the profitable prices has to be managed for the milk producer living in far flung villages.
3. In urban areas milk and milk products have to be supplied at reasonable prices.
4. As consumers do not get pure milk, the production and the milk distribution have to be brought under organized sector, so that the interest of both the consumers and the producers might be protected.
5. Loan has to be granted to the needy milk producers fresh milk free from any sort of adulteration.
6. Villagers have to be imparted a training to procure fresh milk free from any sort of adulteration.
7. To provide first aid famility to cattles in rural areas.
8. Farmers have to be benefited through introducing cattle insurance schemes.

NEED OF INTEGRATED MANAGEMENT SYSTEM IN WHITE REVOLUTION

In addition and this, the main objective of the white revolution is to develop a fully a integrated management system in India that is flexible enough to respond any new management system standard that may be developed. All dairies are composed of a set of systems and process to realize their objectives. Each of these process has layers of subordinate structure with the potential for integration as demonstrated by the following examples :

(i) Vertical Organisation Integration
(ii) Horizontal Integration

In *vertical organisation integration* the broad-based top management policy and strategy is progressively and seamlessly incorporated into a section at every point within the organisation.

Whereas in *Horizontal Integration* where different parts of the dairies are aware of every other parts and all parts are mutually communicative and supportive.

It is also the prime objective to make an effective measure to remove the common concerns of management systems prevailing in the white revolution :

(i) Management process of regional, human resource management, infrastructure and work environment,
(ii) Control of documents and records,
(iii) Understanding customer/stakeholders' requirements,
(iv) Planning for product/service realization process,
(v) Purchase process and suppliers' relationship,
(vi) Control of manufacturing/services realization process to get desired results,
(vii) Preservation of products with efficient storage/warehousing,
(viii) Control of monitoring and measuring devices to ensure consistent results,
(ix) Monitoring and measurement of process of efficient utilization of resources,
(x) Monitoring and measurement of products to ensure conformity,
(xi) Continued improvement control of non-conforming product, corrective and preventive action,
(xii) Control of emission and efficients, and
(xiii) Incident management and recall.

Moreover, these are the systems and processes that come to play for efficient working of the dairy sectors or the white revolution any management system designed has to revolve round these systems and processes.

WHITE REVOLUTION AND MIXED FARMING

White Revolution is a part of *mixed farming* which has many advantages :

(a) well suited for adoption round the year under Indian conditions,
(b) income obtained throughout the year,
(c) offers opportunity for better use of land, capital and labour,

(d) helps in maintaining soil fertility,
(e) reduces the risk due to failure, unfavourable market price, etc.
(f) income in regular and quick,
(g) cost of transportation and sale of by products can be reduced to minimum,
(h) offers opportunity for complete use of industrial wastes, and
(i) provides balanced and protective farming.

FORMS OF FARMING

Farming is categorised under following forms :

(a) *Physical Farming* : (i) Climate, (ii) Social, (iii) Topography.
(b) *Economic Farming* : (i) Marketing cost, (ii) Changes in relative Value of farms products, (iii) Availability of labour and Capital, (iv) Land waste, (v) Cycles of over and underproduction, (vi) miscellaneous seasonal availability of raw material diseases, etc.
(c) *Social Farming* : (i) Type of community, (ii) co-operative spirit.

DAIRY DEVELOPMENT IN INDIA DURING PRE-PLAN PERIOD

Archaeological finds from the Indus Valley Civilization attest the domestication of both cows and buffaloes between 4000 and 3000 B.C. in India. The establishment of cattle breeding farms on the recommendations of the Royal Commission on Agriculture (RCA, 1928), for cattle development formed an important part of the eastern development activities in India. Another important contribution during the late 1930's and early 1940's in the field of cattle development was the laying down of description and characteristics of different breeds of cattle and buffaloes by the Imperial Council of Agricultural Research (ICAR). There were 904 veterinary hospitals and dispensaries in British India at the time of submission of the Report of the RCA.

Whyback in 1930's Wright (1937) while emphasising the importance of improving milk yield in the Indian cattle strongly opposed adoption of large scale crossbreeding to improve the milk yield of the local cattle under the prevailing local conditions and recommended that immediate steps should be taken in selecting and improving milk strains of indigenous cattle. Further Animal Husbandry Wing of the Board of Agriculture and Animal Husbandry of India at its Fourth (1940) and Sixth (1945) NCUI (1952) meeting discussed the question of mass improvement of inferior indigenous cattle and recommended that grading up of cattle should be taken up in a systematic way in selected areas. In 1944 the Government of India invited Papperall for making a rapid survey (Papperall, 1945) of the problem facing dairy industry in the country. He emphasized that milk yield of indigenous breeds should be improved rather than attempting large scale cross-breeding with imported bulls.

Organisation and maintenance of *Gaushalas* (houses for old and unproductive cows) was another important effort made in the country during pre-plan period.

FACTORS REQUIRED FOR WHITE REVOLUTION

The following factors require considerable attention when one decides to go for milk production massive scale farm in terms of white revolution :

(a) suitability of dairy farm,
(b) suitability of the dairy farm, building and other fixed equipments,
(c) supply of right type of labour,
(d) availability of capital,
(e) capability of farmers,
(f) physical condition,
(g) climate, and
(h) water supply

CONDITIONAL BASIS OF DAIRY FARMING/WHITE REVOLUTION

The basis of economic planning of dairy farming depends upon the following :

(i) Dairy farm area devoted to dairy farm and stocking density,
(ii) labour utilization,
(iii) level of milk yield,
(iv) housing facilities,
(v) watching milk yields,
(vi) check on food quantity and quality,
(vii) raising replacement stock,
(viii) seasonal production policy, and
(ix) size of herd.

One should keep in one's mind that the specialization is not always a best policy. Dairy farming is also best profitable in combination with many other enterprises.

(i) Farm Area Devoted to Dairy Farm

The density of stocking (livestock units) is found 50% greater a farm with higher forge productivity and this is mainly due to their higher gross output. The information available (*Indian Dairyman*, Vol. XXXI (3), 1979, p. 188) suggested that dairy unit of 3 cows and followers could be maintained on one acre fertile and fully irrigated land, provided all animals are not purchased at a time.

As to the question of the relatively profitability of using increased forage output to expand the herd or to replace concentrates for the existing herd, the evidence is not clear. It may however, pointed out that the answer to this largely depends on nature of farms, productivity, the level of milk and cost of concentrates.

The fodder grown on the farm must be used to best advantage. Rate of stocking must be so adjusted that food is not wasted, nor the stock underfed. In this connection the following points need considerable attention :

(a) Reserve of feed has to be created for periods of shortage due to unforeseen climatic hazards like drought, flood, lean period of green foddern availability, etc.
(b) Calving date adjustment to secure best results from seasonal fodder crops.

(c) The problem of quantity and quality of fodder supply, since pastures cannot be left long without deterioration.
(d) The use to the best advantage of successful silage or lay making techniques.

(ii) Labour Utilisation

Cost of labour is second to cost the feed in annual cost of keeping a cow. The loose housing system saves labour because cows come to milking parlour instead of man going to cow, manure loader can be used in the loafing area. Cleaning may be done twice a say instead of once a day. They can be self-feed and feeder filled once a day to save labour :

Labour utilization comprises :

(a) milking time/cow/day
(b) dairy chores
(c) Feeding
(d) work.

For labour savings and optimum utilisation, this study shall suggest the following :

(a) proper planning of cattle housing unit for economic labour utilisation,
(b) loose housing system saves labour time and energy,
(c) proper grouping of buildings in layout for saving time and energy of labour is unnecessary movements,
(d) selection of proper equipment which is cheaper, easy to operate durable and give trouble free services, and
(e) good planning of chores for efficient management of animals with utilisation of labour.

(iii) Level of Milk Yield

Statistical evidence appears to favour the high yielding herd. The upward tendency in profit with the increasing milk yield is what one should expect but upto a certain limit only because the food cost per cow also increased due to extra concentrate with the increase in milk yield.

(iv) Housing Facilities

In *white revolution* the vital points to be considered with regard to housing facility are as follows :

(a) The yard and parlour system requires less capital investment per cow and less labour/cow compared to conventional cow shed.
(b) Yard and parlour system needs greater supply of straw for letter.
(c) Grouping of building in a way that facilitates minimum movement of labour to prevent wastage of time in accomplishing daily chores.

(v) Watching Milk Yield

The factors causing variations in milk yield are technical not economical. The annual herd average yield figure is valuable as a guide to the general level of herd efficiency. The dairy milk yield record of an individual cow can be used as guide for rationing, an indication of status of health and faulty feeding end as a basis for culling.

(vi) Check on Food Quantity an Quality

Depending upon the milk yield and requirements of an animal, farmers must work out the ration for each cow and write at down on the chart against the animal. It helps to ensure the supply of right quantities of concentrates, from time to time depending upon quality and quantity of roughages.

With this sound rationing system the dairy farmer should keep a check as quantities of food actually being used either by weekly issue from store or periodical spot checks and seasonal checks.

(vii) Raising Replacement Stock

Most dairy farmers prefer rearing most of their heifers on their farm to maintain required herd strength for two reasons :

(a) To avoid risks of buying poor quality stock or unhealthy animals.

(b) With the introduction of by-products and unconventional feed stuff, it is believed that heifers can be raised more cheaply, than cost of purchasing (Reddy *et. al.* 1984). The calf rearing cost can be minimised by proper breeding scientific feeding, prevention of diseases and judicious management practices.

(viii) Seasonal Production Policy

Milk plants offer incentive in the form of or by way of higher price for milk during lean period of summer months so that the farmers may procure or obtain more milk in those months of high prices. The milk production in summer being relatively more costly its financial advantages are not clear. Further it must be borne in mind that it is not advisable to disturb the calvings unduly to obtain higher milk yields in summer months.

(ix) Size of Herd

According to the findings for National Investigation of milk cows, upto a certain points, herd size has an important influence on the profitability of milk production. No appreciable improvement in profitability was noted with a level of cows above 40. In fact a distinct fall is profits seemed to result above that level. The greater part of variation in profits was found to be due to reduction in costs of labour per cow with the increase in herd size.

The size of herd depends upon following factors :

(a) method of milking
(b) milking shed facilities
(c) milk yield/cow
(d) cow-shed layout
(e) labour efficiency
(f) area under forages.

SIGNIFICANCE OF WHITE REVOLUTION

White Revolution is more significant today as it has following advantages :

1. *Important Human Food*: Milk is palatable, easy to digest and highly nutritious. It is an important human food.
2. *Milk Nearly Perfect Food* : Milk contains fat, milk, sugar, proteins, minerals and liberal source of vitamins.
3. *Milk as a Protective and Balanced Food* : Milk and its product are the only source of animal protein in vegetarian diet. Hence, Nutr. Advisory Committee of ICMR recommended 283 gms/day/per capita to balanced the diet for supply of essential amino acids.
4. Dairying supplies products of worth Rs. 2400 million annually.
5. Sources of draft power for various agricultural operations. Some of the excellent draft breeds supply good quality bullocks—the source of power which brings savings in energy resources like petroleum product and coal.
6. *Dairying suites to Agricultural Operations* : Due to small size holdings of farmers all agricultural operations can best be completed by bullocks.
7. *Dairying provides organic manure* : Dairying provides organic manure which is the best means of maintaining soil fertility and organic farming.
8. Dairying provides opportunity to making use of barren/ unfertile land for housing of animals.
9. Dairying under Indian conditions fits well for agriculture as mixed farming and provides protective and balanced farming.
10. White Revolution offers opportunities of proper utilization of by-products and industrial wastes as cheaper source of feed for animals :

    (i) Utilization of agricultural waste, by-products like wheat *bhusa*, paddy straw, rice polish, wheat bran, cakes, chunnis, etc.

(ii) Utilization of milk by products like whey butter milk for feeding to calves and other growing stock.
(iii) Utilization of animal by-products like bone meal, fish meal, meat meal, blood meal, etc.
(iv) Utilization of industrial by-products like molasses grain godown sweeping, etc.
(v) White Revolution offers opportunity of getting income round the year.
(vi) White revolution contributes a lion share to national income GDP from animal husbandry including contribution from draught power.

There is a greater need of formulating central *Dairy Herd Association* with regional branches consisting of consultants, experts, stockmen, testers. These textures and stockmen should make a periodical check up on all dairy farm for quality and quantity of milk, kinds and quality of feeds given to animals health measures taken, infections diseases reports, etc. These report should be examined by the experts and subsequently by consultant committee. The later would also visit the dairy farm in respect to recommendations made for the improvement mainly through process of culling. A catalogue with identification marks of each animal of different breeds and crossbreeds should be maintained by this association. This would help to standardize the herd management practices to a more or less uniform scale in the country.

HERD RECORDING

In white revolution herd recording occupies an important position. Herd recording originated in Denmark. The first Milk Recording Society was formed at Venjen in Denmark in 1895. Lateron it was adopted by America, Netherland and other European countries. Thereafter, it was adopted by most countries in the world wherever dairying progressed.

METHODS OF MILK RECORDING

Following are the methods of milk recording :

(a) Recording milk daily
(b) Recording milk weekly
(c) Recording milk certain days in a week.

Some dairy farmers record the milk yield of each cow daily. Other prefer to record the milk yield, whereas few prefer to record milk yield, fat percent and quantity of feed given to cows. The accuracy and reliability of data depends upon the method and time interval given for recording milk. For true picture it is better that milk recording be done every day.

ADVANTAGES

White Revolution possesses various gains as it contains following advantages :

(i) Effects of the age of cow, kind of feed, quantity of feed, stage of location, location of length, dry period, calving interval, month and reason of calving and quantity and quality of milk produced can ascertained.
(ii) Pedigree and history sheet can be maintained.
(iii) Some informations like average milk yield, location length, dry period, milk average, herd average can be obtained.
(iv) It is useful for quickest means of improvement (breed-wise) of the herd.
(v) Helps as selection by culling uneconomic animals by knowing persistency and peak yield and average milk of a cow when compared with herd average.
(vi) Helps in fixing the price of animal.
(vii) Animals can be registered in central herd book.
(viii) Helps in economic feeding of animals after determining requirement based on product.

MATERIALS REQUIRED FOR MILK RECORDING

Following materials require for milk recording :

(a) Milk receiving apparatus,
(b) Herd recorder (spring balance) calibrated to weigh minimum of 100 gm. (capacity 10 kg.),

(c) Milk cows,
(d) Strainer, and
(e) Milk produce register.

In the context of livestock revolution white revolution is quite relevant for a developing economy like India. India has a vast resource of white revolution which plays a vital role in the national economy and also in the socio-economic development of rural households. India has the largest population of cattle and buffaloes in the world and accounts for more than 50% of buffaloes and 20% of the cattle population in the world, most of which are milk cows and milk buffaloes.

According to *Indian Dairyman* (February 2002, p. 13) India has a over 20% world bovine buffalo population but only 5% global milk production. Though, we have great population of animals, per animal milk yield including crossbreed. Indian cows and buffaloes is about 800 liters per animal. We have to increase the milk production per animal by well co-ordinated dairy development programme.

The market in India forecasted is based on the following criteria :

The huge middle class in India needs liquid milk. The main bottleneck in liquid milk marketing in India attributed to:

(i) Variable product pricing and distribution strategies.
(ii) Indian household often busy less than 1 liter per day.
(iii) Demand for liquid milk varied from one region to another.
(iv) Seasonal variation in the buffalo base supply affects the less period liquid milk availability.

There is also demand for different type of milk in different cities, e.g. in Calcutta and Bangalore. Cow milk is preferred but some of the sectors like Mumbai and Delhi buffalo milk is preferred. The buffaloes milk is preferred by the majority of Indian consumers. The reason behind this is, the buffalo milk meet the fat demand and produce more solid curd without off flavours.

There is also seasonal variation in the supply of buffalo milk. It has been established that giving rate with 14-16% protein with high energy concentrates, buffalo can give more milk as well as bred regularly.

The seasonal variation can be reduced to 10% by using cow milk. This gap can either filled up by cow milk or by using reconstituted milk. It is generally canvassed to consumers to buy buffalo milk which has more phosphorous, more calcium, solid non-fats and energy.

However, it is often stressed that milk per capita consumption is always measured in terms of total milk consumption and not only liquid milk. Other major protein requirements are fulfilled by pulses and legumes and eggs but the energy source trace elements can be taken from milk only.

Some of the small or big dairy even has no powder plant. Some of the dairies has to take their surplus milk to other dairy and get excess milk converted milk powder. Many dairies exchanging and make contract to buy milk from other dairies private or co-operatives so this linkage has already been established. 23 years back there was a ration card required to buy milk from government dairy in Mumbai and this has changed and milk is available to the consumer throughout the years with some extra price depending upon the demand.

This has been achieved through the public sector organisation, Anand pattern co-operative dairies and private sector.

The problem of adulteration in liquid milk has create serious issue. And it has become a nightmare for liquid milk marketing in the country. We have to evolve strategies to protect us by using tamper proof/tamper evident packaging material such as carton packaging. Packaging materials with holograms, etc.

Thus, under white revolution there is a greater need of tremendous amount of work needed to improve genetics of animals, clean milk production techniques, optimum feed formulation to achieve market driven milk production and distribution. If we achieve these objective of white revolution with collaborative efforts, we shall be able to market more liquid milk and give more returns to our farmers (Table 2.7) demonstrates livestock population.

Following are the turn a round plan adopted for improving financial viability of dairies in India :

A. Milk Procurement

(i) Strees on cow milk given so as to bridge the gap between lean and flush procurement.
(ii) Payment of incentive price, specially during lean season.
(iii) Regular and timely payment to milk produces.
(iv) Increasing poorer members and resultant increase in milk procurement.
(v) Popularising ureas treatment of straw and sale of cattle feed.
(vi) Induction of cross-breeds under various poverty alleviation programmes.
(vii) Consolidating and strengthening functional and revival of difunct DCSs.

B. Marketing

(i) Coverage of area/population per out let.
(ii) Stress on quality assurance programme for better quality of milk and milk products.
(iii) Low cost advertising/awareness creation campaigns.

TABLE 2.7

Livestock Population, All India Census, 1961-97

(*Million*)

Species	*1961*	*1972*	*1982*	*1992*	*1997*
(1)	*(2)*	*(3)*	*(4)*	*(5)*	*(6)*
Cattle	175.60	178.30	192.45	204.58	195.83
Adult female cattle	51.00	53.40	59.21	64.36	NA
Buffalo	51.20	57.40	69.78	81.21	88.71
Adult female buffalo	24.30	28.60	32.50	43.81	NA
Total bovines	226.80	235.70	262.36	289.00	284.54
Sheep	40.20	40.00	48.76	50.78	57.25

(*Contd.*)

TABLE 2.7 (Contd.)

(1)	(2)	(3)	(4)	(5)	(6)
Goat	60.90	67.50	95.25	115.28	121.40
Horses and Ponies	1.30	0.90	0.90	0.82	0.59
Camels	0.90	1.10	1.08	1.03	0.89
Pigs	5.20	6.90	10.07	12.79	13.38
Mules	0.05	0.08	0.13	0.19	0.11
Donkeys	1.10	1.00	1.02	0.97	0.61
Yak	0.02	0.04	0.13	0.06	0.06
Total livestock	335.40	353.60	419.59	470.86	478.85
Poultry	114.20	138.50	207.74	307.07	357.13
Dogs	NA	NA	18.54	21.77	NA

Source : Department of Animal Husbandry & Dairying, Ministry of Agriculture, Government of India.

TABLE 2.8
Average Composition of Milk of different Animal Species

(*Percent*)

Species	*Water*	*Fat*	*Protein*	*Lactose*	*Ash*
Friesian cow	87.92	3.40	3.13	4.88	0.69
Sindhi cow	88.07	4.90	3.42	4.91	0.70
Gir cow	88.44	4.73	3.32	4.85	0.66
Tharparkar cow	86.58	4.55	3.36	4.83	0.68
Sahiwal cow	86.42	4.55	3.33	5.04	0.66
Crossbred cow	86.54	4.50	3.37	4.92	0.67
Buffalo	82.76	7.38	3.60	5.48	0.78
Goat	87.10	4.25	3.52	4.27	0.86
Sheep	81.00	7.90	5.80	4.50	0.80
Camel	86.50	3.10	4.00	5.60	0.80

Source : *Ibid.*

(iv) Better services to the retailers and consumers.
(v) Production and marketing of value added/fresh milk products.

C. Critical Financial Parameters

(i) Regular monitoring of various operational costs and adoption of measures to reduce the same.

(ii) Effective use of installed capacities of plants by improved through out.

FACTORS INFLUENCING MILK COMPOSITION

Bacterial growth in milk can be retarded by refrigeration, thereby slowing down the rate of its deterioration. Under certain conditions, refrigeration may not be feasible due to economical and/or technical reasons. In such situations, another method has been developed for retarding bacterial growth in raw milk during collection and transportation to dairy plants.

The traditional practice in India of boiling the raw milk within hours of production before its consumption provides a safety net for minimizing occurrence of microbes. However, pasteurization is designed to destroy relatively more heat resistant pathogens.

Milk is valued commercial for its two important parameters : (i) Milk Fat (F), and (ii) Solids Not–Fat (SNF). Total Solids (TS) refers to the quantity of SNF plus fat present in milk. It may range from 12 to 16 percent, depending on its source. For cow milk, TS is 12 percent (3.5% F and 8.5% SNF) while for buffalo milk it ranges between 15 and 16 percent (6-7% F and 9% SNF).

Apart from breed related differences certain other factors also influence the gross composition of milk. They are :

- individuality of animal;
- state of milking;
- intervals of milking;
- completeness of milking;
- frequency of milking;
- irregularity of milking;
- portion of milking;
- different quarters of udder;
- lactation period;
- yield of milk;
- season;
- feed;
- nutritional level;

- environmental temperature;
- health status;
- age;
- weather;
- oestrum or heat;
- gestation period;
- exercise;
- excitement; and
- administration of drugs and hormones.

In general, these variables tend to average out but show a seasonal pattern in the production of commercial milk used by dairy processors.

Suitability of Cow/Buffalo Milk for Products' Manufacture

In India, commercial milk is mostly a combination of buffalo and cow milk. Therefore, it is important to recognize both qualitative and quantative differences in the major constituents of buffalo and cow milk in Table 2.9.

TABLE 2.9
Composition of Cow and Buffalo Milk

(per cent)

Constituent	*Cow Milk*	*Buffalo Milk*
Water	86.50	83.18
Fat	4.39	6.71
Protein	3.30	4.52
Lactose	4.44	4.45
Total solids	13.50	16.82
Solids-not-fat	9.11	10.11
Ash	0.73	0.80
Calcium	0.12	0.18
Magnesium	0.01	0.02
Sodium	0.05	0.04
Potassium	0.15	0.11
Phosphorus	0.10	0.10
Citrate	0.18	0.18
Chloride	0.10	0.07
Calcium/Phosphorus ratio	1.20	1.80

TABLE 2.10

Import and Export of Milk and Milk Products, 1995-99

(Quantity in tonnes and value in Rs. million)

Products	*1995-96*		*1997-98*		*1999-2000*	
	Quantity	*Value*	*Quantity*	*Value*	*Quantity*	*Value*
IMPORTS						
Milk and cream concentrated/containing sugar/sweeting matter	5,013.4	366.3	751.3	51.5	17,712.7	1,037.7
Whey and products concentrated of natural milk constituents	78.5	3.8	6.0	0.3	447.3	28.6
Butter and other fats and oils derived from milk	4,010.1	249.0	4,352.1	238.9	10,246.2	696.5
Cheese and Curd	72.3	9.6	46.9	6.8	329.0	34.0
EXPORTS						
Milk and cream concentrated/containing sugar/sweeting matter					80.3	5.8
Butter milk, cream, yogurt, etc.	3,590.5	195.1	1,606.9	95.9	4,291.9	292.8
Butter and other fats and oil derived from milk		1.5	25.2	0.7	9.7	0.4
Cheese and curd	563.1	66.0	297.8	33.3	1,639.3	186.5

Source : Director General of Commercial Intelligence and Statistics, Kolkata.

TABLE 2.11
PFA Standards for Different Classes and Designations of Milk in India

Class of Milk	Designation	State and Union Territories	Millenium %	
			Milk fat	Milk Solids not-fat (SNF)
(1)	(2)	(3)	(4)	(5)
Buffalo Milk	Raw, pasteruized boiled, flavoured and sterlized	Assam, Bihar, Chandigarh, Delhi, Gujarat, Maharashtra, Haryana, Meghalaya, Punjab, Sikkim, U.P., West Bengal, Andaman & Nicobar, Andhra Pradeesh, Arunachal Pradesh, Dadra & Nagar Haveli, Goa, Daman & Diu, Kerala, Himachal Pradesh, Jammu & Kashmir, Karnataka	6.0	9.0
	-do-	Kerala, Lakshadweep, Madhya Pradesh, Manipur, Mizoram, Nagaland, Orissa, Pondicherry, Rajasthan, Tripura, Tamil Nadu.	5.0	9.0
Cow Milk	-do-	Chandigarh, Haryana, Punjab	4.0	8.5
		Andaman & Nicobar, Andhra Pradesh, Arunachal Pradesh, Assam, Bihar, Dadra and Nagar Haveli, Delhi, Goa, Daman & Diu, Gujarat, Himachal Pradesh, Jammu & Kashmir, Karnataka, Kerala, Lakshadweep, Madhya Pradesh, Maharashtra, Manipur, Meghalaya, Nagaland, Pondicherry, Rajasthan, Sikkim, Tamil Nadu, Uttar Pradesh, West Bengal	3.5	8.5
	-do-	Mizoram, Orissa	3.0	8.5

Goat or Sheep Milk	-do-	Chandigarh, Haryana, Kerala, Madhya Pradesh, Maharashtra, Punjab, Uttar Pradesh	3.0	9.0
		Andaman & Nicobar, Andhra Pradesh, Arunachal Pradesh, Assam, Bihar, Dadra & Nagar Haveli, Delhi, Goa, Daman & Diu, Gujarat, Himachal Pradesh, Jammu & Kashmir, Karnataka, Lakshdweep, Manipur, Meghalaya, Mizoram, Nagaland, Orissa, Pondicherry, Rajasthan, Sikkim,	3.5	9.0
		Tamil Nadu, Tripura, West Bengal	3.0	9.0
Mixed Milk	-do-	All India	4.5	8.5
Standardized Milk	Pasteurized, flavoured and sterlized	All India	4.5	8.5
Toned Milk	-do-	All India	3.0	8.5
Double Toned Milk	-do-	All India	1.5	9.0
Skimmed Milk	-do-	All India	Not more than 0.5	8.7
Full cream milk	Pasteurized and sterlized	All India	6.0	9.0

Source : Prevention of Food Adulteration (PFA) Rules, 1995.

Physical Equalibria

Milk has well-defined physical equilibria among its various constituents that exist mainly in three forms—as an emulsion, as colloids and in true solution. In milk, fat is present as an emulsion, protein and some mineral matters in colloidal suspension, and lactose together with some minerals and soluble proteins in true solution.

Notes and References

1. James, N. Warner (1953) : *'Dairying in India'*, Macmillan and Company Ltd., London.
2. Peterson, W.F. & Field, A.M. (1954) : *'Dairy Farming'*, J.B. Lippincott Company, New York.
3. Buruns, A.R. (1955) : *'Comparative Economic Organisation'*, Prentice Hall, New York.
4. USDA (1999) : *'Implementation of Uruguay Round Tariff Reduction', Agricultural Outlook*, November, U.S. Department of Agriculture, Washington, D.C.
5. USDA, FAS (2000) : *'Dairy : World Markets and Trade'*, July (Edition), United State, Dept. of Agriculture, Foreign Agriculture Service, Washington, D.C.
6. WTO (1995) : *'The World Dairy Markets for Dairy Products'*, No. 95-3814, Geneva, The World Trade Organisation.
7. Banerjee, A. (1999) : *'Expert Potential of Indian Dairy Products', Indian Dairyman*, Vol. 50, No. 4, pp. 19-22.
8. Parthasarthy, S. (2000) : *'National Policies Supporting Small holder Dairy Production and Marketing in India'*, Paper presented at the NDDB–ILRI South. South Workshop on Smallholder Dairy Production and Marketing. Constraints and Opportunities, National Dairy Development Board (NDDB), Anand, March 13-16, 2001.
9. National Dairy Development Board (1965) : NDDB, Anand, p. 24.
10. John, P. (1975) : *'Economic of Dairy Development in India'*, Prabhat Prakashan, Digha Ghat, Patna.
11. Rajorhia, G.S. (1991) : Evaluation of the Quality of Khoa Prepared from Different Mechanized System, *Indian Journal of Dairy Science*, 44 (2), 181-87.
12. Jain, S.C. *et. al.* (1990) :
13. Chennegowda, H. (2002): *'Liquid Milk Marketing—Opportunities and Threats', Indian Dairyman*, No. 54 (February) pp. 99-102.
14. Prashan, K. (2001) : *Indian Dairyman*, 53, 10, 2001, p. 34.

References

Banerjee, A.K. (1996) : *Indian Dairy : An Overview*, Productivity, Vol. 36, No. 4.

Cluruc Harvey & Harvy Hill (1951) : *Milk Production & Control*, M.N.K. Lewis & Co., London.

Das, G.S. & D.S. Sidhu (1973) : A study of Milk Production and Supply Function in Ludhiana District, Punjab, 12th Annual Report, IAU, Ludhiana.

Elwood, M.J. (1964) : *Approved Practice as Dairying*, Oxford & PBH Publishing Company, New Delhi.

Fine, J.C. *et. al.* (1982) : *Livestock in Asia*, Issues & Policies, IIDC, London.

Giridhari, D.G. (1982) : *'Dairy Development'*, Indian Journal of Marketing.

George, P.S. & V.K. Shrivastava (1975) : *Dairy Development : Decision at Farm Level*, CMA Monograph No. 66, IIM, Ahmedabad.

Hall, C.W. (1971) : *Dairying of Milk & Milk Products : The AVI*, Publishing Company Webtport, Connectient.

Hall, H.S. (2000) : *Milk Plant Layout*, FAO Agriculture Studies Bulletin No. 39, WHO/FAO, Rome, Italy.

Mishra, S.N. (1979) : *Livestock Planning in India*, Vikas Publishing House, New Delhi.

Nair, K.N. : *White Resolution in India : Facts and Issues*, Economic and Political Weekly, 20 (25 & 26) A 89-A 95.

Ramgappe, K.S., Acharya, K.T. (1973) : *Indian Dairy Product*, Asia Publishing House, Bombay, pp. 149-363.

Rao, V.M. (1980) : *Towards White Revolution*, Kisan World, Vol. 10, No. 8, pp. 26-27.

Ravindra, D. (2001) : *Indian Dairy Industry : Present Scenario & Future Investment Potential*, Food & Pack, Vol. No. 2, No. 7.

Singh, Kartar (1998) : *Dairy Development in India, Retrospect and Prospect*, IRMA Anand, November, 1998.

3

Role of Dairy Farming in Economic Development

The *'dairy farming'* or the *'White Revolution'* in our country is developing at a rapid and accelerated rate. The indigenous milk products are manufactured by adopting unit operations. The traditional milk products are the products of masses being made in India since time immemorial. These products have great social religious, cultural, medicinal and economic importance. It is estimated that more than half of 78 million tonnes of milk produced in India is utilized for the manufacture of traditional milk products. Indian dairy is one of the most modern and sophisticated dairy industry in the world producing huge quantum of milk and milk products like: butter, ghee, paneer, khoa, chana, chakka, curd, srikhand in addition to pasteurisation of milk, standardisation and dispatching to the consumers.[1] According to B.M. Vyas, Ex-Managing Director of Gujarat Co-operative Milk Marketing Federation, Anand.[2] "Dairy foods form an important constituent of the daily menu for people world over the difference in proportion being positively correlated with the relative affluence of the society. Not surprising at all, given that they form a balanced food item one that lends itself to exquisite culinary menoeuvres."

Indian dairy industry has made rapid progress since independence. A large number of modern milk plants have since been established.

India has emerged as the largest producer of milk in the world with a total production about 100 million tonnes in 2005 followed by United States with milk output of 95 million tonnes (FAO, 2005). In the global context, the performance of Indian dairy sector appears impressive in terms of livestock population and total milk production, but very poor in terms of productivity. The dairy sector has been playing a very significant role in improving the economic condition of our rural population. A majority of the rural households of the country own livestocks. Our farmers have benefited from different dairy development schemes and India today has the distinction of being the highest produce of milk the world with an average growth rate between 4-6% per annum, the credit for which largely goes to 10 million farmers associated with dairy farming through more than 82,000 village Dairy Co-operative Societies.

Traditional Milk products represent the most prolific segment of our India Dairy Industry. Despite the immensity of volume of milk handled, preparation and marketing are confined to the unorganised sector.[3]

Since most of the Western type dairy products manufactured by the unorganised sector of the Dairy Industry are reaching near saturation level in the existing domestic and international markets, the entire range of Indian milk products represent the most promising venue for diversification.

Furthermore, dairying has played a prominent role in strengthening our rural economy. It has been recognized as an instrument to bring socio-economic transformation by helping the landless and marginal farmers.[4]

Access to modern technologies, research, development and creation of infrastructural facilities has given tremendous boost to dairy development specially in the co-operative sector. In addition to employment generation amongst the rural people, dairying has helped in improving the quality of life of the rural masses.

For sustaining further development, Indian dairy sector would have to cope with the rapid transformations that are taking place in the world economies, consequent to the GATT Agreement as global trade of dairy products are being strictly guided and regulated by the WTO guidelines under the provisions of AOA in terms of : (i) access of dairying, (ii) domestic support to dairy sector, (iii) export competition of dairy product. Newer and stricter sanitary and phytosanitary standards are being formed for regulating quality standards and parameters of the export of dairy products. Under these newly emerging circumstances, quality standards for production and processing milk can't remain at variance with the global standards.

Indian Dairy product have not only served as a cultural link with the modern dairy industry but also provide technological base for diversification, export promotion and a value added product to make the modern dairy sector economically strong to enable the milk producer to benefit from it.

Today, Indian milk products are the largest and fastest growing segment of the dairy industry. Dairy products offer opportunities for absorbing the growing milk surplus, generated by the *Operation Flood*. In recent years, the focus of R and D has been on the application of technology for the mass production of indigenous sweets.

DAIRY STRUCTURE

The structure of the dairy in developed countries is characterised by high per capita consumption and slow growth rate with regard to Asian Countries with lower per capita consumption, poised to grow at a fast rate. Consumer preferences change with education improved economic conditions, changing aspirations and cultural orientation. Markets undergo gradual and rapid changes involving via complex interaction with technology and socio-politico-economic drivers.[5]

The world is shrinking. Distances are collapsing because of instantaneous communication and faster transport. Technology is changing the way of the world. Information and communication would bring people together live never before. All barriers to movement of goods, services world view, cultures, spiritual enables are targeted demolition. Broadly speaking, demographics, income and taste preferences drive the demand for food items true milk products. The objectives of this exercise is to ascertain the current status of these variables, viz. population pattern, income structure, life style and the way there variables shape the demand for handful of the dairy products in Asia and the developed world. The discussion beguns with demand drivers and the way they develop.[6]

It is expected that the population of Asia will grow at 1.08 percent between 1997 and 2020, while the developed world will stangnate. In 2020 Asia's population growth will be three times the population of developed world. The higher population growth would augment the dairy demand growth rate in Asia, while the stagnation of population is developed world lead to flat dairy demand.

The importance of livestock is general and dairying in particular hardly needs emphasis in a country like India. India has vast resources of livestock, which play a vital role in the national economy and also in the socio-economic development of rural households. India has the largest population of cattle and buffaloes in the world. Indian dairy sector contributes a large share to the agricultural Gross Domestic Product (GDP). Milk and milk products constitute a major share in the value of output from dairy sector.

In India, a large proportion of population is vegetarian, in there diet milk and milk product assume great importance, as they are the only source of animal protein in their diet. Here, both cow and buffaloes are important animals for producing milk. The milk of goat ship, and to a lesser extent that of camel and ass is also used. Goat milk is used in most parts of the country, continuing about 4% of the country's total production, as against about 53% provided by the buffaloes and 33% by the cow.

HERITAGE OF INDIAN MILK PRODUCTS

Most people eat dairy products for pleasure. It dictates all consumer demands. It implies something different, a shift in habits to satisfy the consumer's desire for pleasure, three new trends are emerging as for as taste is concerned in popular dairy product segments : (a) Authenticity, (b) Variety, (c) Exoticims.

Of late, dairymen have realized the need to change the image of milk as a health/nutritive product that people must consume to a mod image of like to for pure pleasure. This would be possible by widening the choice of dairy products and adding into then authentic, exotic flavours. There dairy products are part of the world heritage and embody the tradition of excellence. They are being reinvested by applying modern process technologies for mass production to meet the consumer's new demands in response to changing lifestyle that combines purity, quality and convenience.

The flavour of new millennium in India's ethnic milk-based sweets, deserts and pudding. Each product has distinctive wisdom. Milk and Milk products are highly valued in Indian societies as a source of nutrition.

In India, milk sweets are an inseparable part of wedding ceremonies, feasts, festival and social and religious occasions. Boxes of sweets are a harbinger of good news, be it a birth or betro that in the family, anniversaries, success in the examination, landing the first job, promotion and the like.

The products developed, either for direct consumption or as an intermediate base material have followed age-old methods of preservation and conservation through, heat desiccation, fermentation, coagulation and classification with an in the recover the total solids to the maximum extent through tiny scale/somehold level processes and technologies that are adequate in total situation. Thus, from time immemorial, milk and milk products have enjoyed eminent position in the Indian food ethos.

Dairy farming in India is still not so developed as it is in the country like Denmark, Sweden, Canada, Australia, New Zealand, USA, etc. In India Dairy farming began in 1881 when certain separators were first introduced. The first large scale dairy farm was started by the military in 1891, at Allahabad and the development of more dairy farm led to the creation of the post-Indian Dairy expert in 1920. Its growth of quite slow before Independence.[7]

But after the Independence Government of India tried to pay attention towards the significance of dairy farming in India.

During First Five Year Plan some steps were initiated. However, it developed very fastly when Green Revolution was launched by Mrs. Indira Gandhi and earlier slogan given by late Prime Minister Shri Lal Bahadur Shastri as 'जय जवान जय किसान'. She announced 'White Revolution' for rapid and accelerated development of dairy farming in India.

With the help of the World Food Programme (WFP) the Department of Agriculture, Government of India, formulated a project for stimulating milk marketing and dairy development in India under *Operation Food Project*. WFP agreed to supply free of cost during Five years period for 1970-71 to 1974-75, 125 thousand tonnes of skimmed milk powder and 42 thousand tonnes of butter oil, or worth of Rs. 41.90 crores the international price. After recombination of the skimmed milk powder and butter oil into liquid milk at the public sector dairies at Mumbai, Kolkata, New Delhi and Chennai, the milk would be sold and the sale proceeds from the quantity estimated at Rs. 95.20 crore would be used for increasing milk processing facilities of the public sector dairies from 1.00 million litres at present to 2.75 million litres per day at the end of five year period.

DAIRY OUTPUT MULTIPLIER

The annual data of milk production for the period 1989-2004 are used to project milk production in the year 2005-10. Following the approach of International Food Policy Research, Washington (IFPRI, 1990), a semi-logarithmic trend equation is fitted to the annual production date to obtain growth rate per annum.

The major factor usually influences the growth in consumption of milk are : (a) Growth rate is population, (b) Growth rate is real per capita income, (c) Income elasticity of demand for milk, (d) Changes in real prices of milk and other closely related goods, (e) Change in the preference of consumers.

Moreover, it could be postulated that the supply side forces would be stronger in coming days, the shortfall would still be unavailable. This would otherwise suggest that the rate of growth is milk production needs to be considerably augmented to catch up with the anticipated demand.[8]

The other option could be a phased reduction is consumption multiplier, so that demand and supply parity could be established. However, this is not easy task. The Bihar State has a strong cultural heritage consume proportionately higher quality of milk and milk products, incidence of milk consumption, is highest among the households and a relatively large population following vegetarian food habits.

It is also significant that the average per capita income, level of urbanisation in the state is relatively low.

The milk supply multiplier in Bihar and other states has been tandy. This is indeed a major challenge before community. But the production augmentation through increase in animal population suffers from inherent limitation—agricultural growth and availability of farm level residues for livestock in unequitably balanced in many states.

Admittedly, the environmental constraints are natural. Thus, a selective area specific approach needs to be formulated for intensive breed upgradation and continuously monitoring the programme apparently, the structural change in composition of milk animal is a boon for the state.

Moreover, it is found that during the more recent years the growth in productivity of the buffaloes has been marginally higher, compared to cattle. The lactation efficiency of the cattle also leads to an identical effect.

The growth in milch animal population is positive. The absolute growth is milch population is found to be inadequate in relation to growth in human population. Thus, the need for enhancing productivity per unit of animal is absolutely unavoidable. The incidence of cross-breeding of local cows with the exotic breed is rather limited. The urban concentration of the cross-breeds is conspirous. Perhaps, it is not a desirable situation for long-term sustainability.

In the dairy sector, there is a constant pressure on limited animal resources to cater to an ever-increasing demographic pressure.[9]

ROLE OF DAIRY FARMING DURING THE POST-ECONOMIC REFORMS

During post-economic reform period i.e. since 1991 dairy farming has been growing quite rapidly. At present, milk product plants are working at Ahmedabad, Anand, Aligarh, Baroda, Bangalore, Coimbatore, Chandigarh, Mehsana, Amritsar, Jind, Moradabad, Kolhapur, Hubli, Dharwar, Pondicherry, Hyderabad, Thiruvananthapuram, Kanyakumari, Rajkot and Vijayawada, Barauni, Patna, Gaya, Bhagalpur, Ranchi, Jamshedpur and Dhanbad.

The Major Dairy Centres in India

States		*Centres of Dairy Industry*
Assam	:	Guahati, Shillong.
Bihar	:	Patna, Bhagalpur, Gaya, Darbhanga, Muzaffarpur, Motihari, Barauni, Hazipur, Madhubani
Jharkhand	:	Dhanbad, Ranchi, Jamshedpur.
Gujarat	:	Ahmedabad, Baroda, Anand, Rajkot, Bhawnagar, Jamnagar, Junagarh, Surat, Mehsana, Surendranagar, Gandhinagar.
Andhra Pradesh	:	Chittor, Guntur, Hyderabad, Nellore, Rajamundary, Vishakhapattanam, Vijayawada, Warangal
Delhi	:	New Delhi
Karnataka	:	Bangalore, Belgaun, Gulberga, Mysore, Mangalore, Devangarh, Bharavati, Hubli-Dharwar.
Haryana	:	Ambala, Jind, Bhiwani.
Rajasthan	:	Jaipur, Jodhpur, Udaipur, Bikaner, Alwer, Bharatpur.
Punjab	:	Amritsar, Ludhiana, Jind, Jalandhar, Hissar, Dasuya, Gohna.
Kerala	:	Thiruvanathpuram, Calicut, Kottayam, Cannanore, Alleppy, Palghat, Ernakulam.

Madhya Pradesh	:	Indore, Bhopal, Gwalior, Jabalpur, Ujjain, Sagar, Khandawa, Guna, Rewa, Katni, Bind, Ratlam, Chindwara.
Chhatisgarh	:	Raipur, Bilaspur, Durg.
Orissa	:	Bhubaneshwar, Cuttak, Puri.
Maharashtra	:	Bombay (Mumbai—Arey), Aurangabad, Akola, Amrawati, Ahmednagar, Dharwar, Dhulia, Nagpur, Nasik, Pune, Solapur, Kolhapur, Sanghali, Miraj, Udaygir, Kudigee.
Tamil Nadu	:	Chennai, Coimbatore, Ootacamund, Chidambaram, Kanyakumari, Kodai Kanal, Tiruchirapalli, Thanjavur, Madurai, Erode.
West Bengal	:	Calcutta, Hoogly, Heringhata.

DAIRY FARMING AND ECONOMIC DEVELOPMENT

The importance of Dairy farming in the economic development of any country whether rich or poor, is borne out by the fact that it is the primary sector of the economy which provides the basic ingredients necessary for the existence of mankind and also provides most of the raw materials, which when transformed into finished products, serve as basic necessities of the human race. Dairy farming plays a most strategic role from several point of view. At a minimum, dairy products must be increased rapidly enough to keep pace with population growth. However, in a speedy industrialising economy, this is not enough. Industrialisation necessarily brings with it urbanisation and a rapid expansion of the industrial labour force. This may, then be expected to bring with it a rising per capita of food and milk and dairy product based on higher urban income. In addition, dairy farming must generate export surpluses in order to earn the foreign exchange with which to finance the import of capital goods and certain kinds of industrial raw materials.[10]

DAIRY SECTOR AS A BACKBONE

Today, dairy sector has become the backbone of an economy which provides the basic ingredients to mankind and raw materials to industrialisation. The role of dairy farming for development of an economy may be attributed in following manner:[11]

(i) Contribution to national income
(ii) Necessary for food shifts
(iii) Pre-requisite for raw materials
(iv) Provision of surplus
(v) Shift of manpower
(vi) Creation of infrastructure
(vii) Helpful to reduce inequality
(viii) Based on democratic nations
(ix) Create effective demand.

FOSTERING ECONOMIC ADVANCEMENT

The lesson drawn from the economic history of many advanced countries tell us that prosperity of White Revolution/ dairy farming contributed to fostering economic advancement. The development of dairy farming is necessary to meet the increased demand of food stuffs in under developed economies. The demand for dairy food is bound to increase with the increase in population. In developing economies, population rises because of a greater decline in the death rate. Some of the under-developed economies of the world are facing population explosion, so the progress of dairy farming should be achieved at a faster rate to cope up with the demand for dairy food otherwise a tremendous in the process of dairy food.[12]

IMPROVEMENT OF SUPPLY OF RAW-MATERIALS

The advancement of dairy sector is necessary for improving the supply of raw-materials for dairy based industries especially in developing countries like India. The shortage of agricultural goods and dairy products has its impact upon on industrial production, and consequent increase in the general price level. It will impede the growth of the country's economy.

INCREASING EXPORT OF DAIRY-PRODUCTS

The progress in dairy sector provides for increasing the exports of dairy products. In the earlier stage of development, an increase in the export earning is more desirable because of the greater strains on the foreign exchange situation needed for the functioning of imports of basic and essential capital goods. Johnson and Mellor are of the opinion, "In view of the urgent need for enlarged foreign exchange earnings and the lack of alternative opportunities, substantial expansion of dairy export production is frequently a rational policy even though the world supply demand situation for a commodity is unfavourable."

ABSORPTION OF HUGE LABOUR FORCE

Dairy farming also absorbs a large quantity of labour force. Dairy progress pursuits the shift of manpower from dairy to non-dairy sector. In the initial stages, the diversion of labour from dairy to non-dairy sector is more significant from the point of view of economic development as it cases the burden of surplus labour force over the limited land. Thus, the release of surplus manpower from the dairy farming sector is necessary for the progress of dairy sector and for expanding the non-dairy sector. The development of dairy farming sector requires road, market yard, storage transportation railways, postal services and many other infrastructure for creating demand of industrial products and the development of commercial sector.[13]

The development of dairy sector has minimised the burden of several developed economies who were facing the shortage of foreign capital. If foreign capital is available with the 'strings' attached to it, it will create another significant problem. Dairy sector requires less capital for its development thus it minimises growth problem of foreign capital.[14]

REDUCTION OF INCOME INEQUALITY

In India, like other predominantly agricultural overpopulated countries, there is a greater inequality of income in the rural and urban areas of the economy. To reduce inequality of income, it is necessary to accord higher priority to dairy farming/third revolution or second wave of White Revolution. The prosperity of dairy farming would raise the income of majority of the rural population and thus, the income disparity may be reduced to a certain extent.[15]

PROMOTE EFFECTIVE DEMAND

Dairy farming also promotes effective demand. The development of dairy farming would tend to increase the purchasing power of dairy farmers which will help the growth of non-dairy and non-agricultural sector of the economy.[16] It will provide a market for increased production. In under-developed countries, it is well known that majority of people depend upon agriculture and it is they who must be able to afford to consume the goods produced. Thus, it will be helpful in stimulating the growth of the non-agricultural and non-dairy sector. Improvement in the productivity of dairy products may pave the way for promotion of exchange economy, which may help the growth of non-dairy sector.

IMPACT OF DAIRY DEVELOPMENT

The impact of dairy development on nutrition and income has been a matter of great interest to policy-makers. At the same time it has been a subject of fierce controversy both in popular press as well as academic journals. The proponents of the dairy development programme feel that such activity does indeed raise the level of income and hence, the nutrition of the rural poor. As such, the dairy development programme needs to be encouraged through the positive policy of the government. (Review of Agriculture, March 28, 1987).

In contrast, its critics assert that speed of dairying in rural areas in leading to transfer of items of nutrition from rural poor to the well to do is the urban areas. Even otherwise the contribution of dairy development programmes to income is rural areas is marginal :

(i) The dairy development has positively contributed towards the improvement of the economic and nutritional status of the rural areas.
(ii) The lower income groups in rural areas derived positive benefits from dairy development in terms of food intake and income.
(iii) There has been a shortage of fodder in rural areas with dairy development programmes because of lack of integration of dairy development with other economic activities needed to support dairying.

Milk and the products are traditional and highly acceptable form of animal products in India. A great deal has of course been done for dairy development and considerable investments have been made. Apart from its great popularity and acceptability is India milk is known to be a nutritive food.

IMPACT OF DAIRY DEVELOPMENT AND NUTRITION

The impact of dairy development on nutrition and income has been a matter of great interest to policy-makers. At the same time, it has been a subject of force controversy both in popular press as well as academic journals. The proponents of the dairy development programme feel that such activity does indeed raise the level of income and hence, the nutrition of the rural poor.[17]

As such, dairy development programme needs to be encouraged through positive governments policy. In contrast its critics assert that spread of dairy in rural areas is leading to transfer of items of nutrition from the rural poor to the well-to-do in the urban areas. Even otherwise the contribution of dairy development programmes to incomes in rural areas is marginal.

The dairy development has positively contributed towards the improvement of the economic and nutritional status of the rural areas. The lower income groups in rural areas desired positive benefits from the dairy development in terms of food intake and income.

There has been a shortage of foddar in rural areas with dairy development programmes because of lack of integration of dairy development with other economic activities needed to support dairying.

To identify whether dairy development has any positive influence on economic and nutritional level of rural population food intake has been dealt in four income groups based on land holding sizes.

The households of villages having dairy development programmes the food intake is higher compared to that of his households of village without dairy development programmes. The average calories and protein intake is also higher in dairy villages. The improvement in nutritional status comes from the fact that milk protein has a higher protein efficiency ratio. Dairy development programmes provide a continuous source of income, the agricultural employment is low.

INPUT-OUTPUT RELATIONSHIP IN DAIRY FARMING

In an dairy production activity, the most fundamental aspects is to examine the various inputs required and the relationship of different levels of each input with the dairy output. Without such an information, the business planning of dairy product to rationally, allocate resources of the production of different dairy enterprises cannot take place. In dairy sector such an analysis is of greater significance due to peculiar nature of business itself. The variety of input required is very large and the response of biological materials like animal (live stocks) to the input use is highly variable due to their explosive to a variety of risks such as climatic factors etc.[18]

The factor product relationship or input-output relationship in dairy farming depends upon :

(A) Technological relationship, i.e. the levels of output at various levels of input in dairy farming.
(B) The prices of the inputs and product prices.

The technical relationship in dairy farming are determined not only by the dairy input used but also by a number of other factors which include the time and methods of their application, climatic factors and other associated institutional factors. Similarly, the availability of the resources, size of dairy farming, strength of labour union, transport facilities, etc. have some influence on the input-output relationships in dairy sector.[19]

Thus, the resources used in the dairy production, process can be categorised as :

1. Stock in one year.
2. Stock over longer period but flow in each year.
3. Flow forever.

PRODUCTION RELATIONSHIP IN DAIRY FARMING

Production relationship in dairy farming can be categorised in following three forms :[20]

(A) Factor-Factor Relationship in Dairy Farming

Various combination of input can be used in dairy farming for producing a certain term of output. The change in the use of a dairy input with other inputs to produce a given level of dairy product is studied under factor-factor relationship

$$x_1 = f(x_2, x_3 ...)$$

Input (x_1) is a function of inputs (x_2, x_3...) keeping the level of dairy output (y) as constant.

(B) Product-Product Relationship in Dairy Farming

Various dairy enterprises are generally related to each other in terms of use and availability of dairy resources. Thus, they may help each other or may cause hindrance to each other in dairy productions. For example, dairying and crop farming help each other in production viz., dairying provides farm yard manure which is useful in crop production. The crop farm helps dairying in providing feed and fodder at cheaper rates. It can be shown in following formulae :

$$y_1 = f(y_2, y_3 ...)$$

The enterprise (y_1) is dependent upon the level of enterprises ($y_2, y_3...$). The relationship may be complementary, supplementary or competitive.

(C) Factor-Product Relationship in Dairy Farming

Most of the production decisions at dairy farm level and at macro level are based on this type of relationship in dairy farming. The level of dairy input required to produce a certain level of output or the output realized at varying levels of inputs is commonly called as factor, product or input-output relationship in dairy farming. It can be explained as:

$$y = f(x_1, x_2, x_3 ... y_n)$$

y is the function of or dependent upon or varies with or related to x_1

y = Output of dairy farming
f = Function
x = Inputs of dairy farming as ($x_1, x_2, x_3 ... x_n$)

In practice, production of a dairy enterprise is not a function of only one input but a number of inputs. Thus, the production function in dairying involving a number of independent input factors is generally denoted as :

$$y = f(x_1, x_2, x_3 ... x_n)$$

where y is a function of the n factors $x_1, x_2, x_3 ... x_n$. If our interest is to find out the effect of a few factors on production, we can write it as $y = f(x_1, x_2, x_3, x_4, x_5)$. This shows that y is a function of x_1, x_2, x_3 while x_y and x_5 are held constant in dairying.

FORMS OF PRODUCTION FUNCTION IN DAIRY FARMING

The relationship between output and input in dairy farming can take either of three forms :

1. Constant Return in Dairy Farming

The constant return or constant productivity hold true if all the units of input applied is production result in equal addition to the total output of dairy product.

As obvious from Table 3.1, through this type of situation where the constant return operates in dairy sector is limited over a small range but the situation is quite common at the lower levels input use.

TABLE 3.1

Input of Dairy Farming	*Output of Dairy Farming*
0	5
10	10
20	15
30	20
40	25

Over the range of its application, it pays either to produce indefinitely or not to produce at all. In case the input-output price indicator suggests that it is profitable to produce one unit of dairy product, then it is also profitable to produce to the extent linear relationship exists. But if it is not advisable to produce first unit, it is also advisable not to produce any more unit on linear scale of product function.

2. Increasing Return Function in Dairying

If every successive increase in the input results in higher and higher additional output, the increasing return function is said to be in operation. It can be illustrated in Table 3.2 as well.

Since, the marginal physical output in dairy farming is increasing throughout, it is desirable not to stop dairy production over the range of this curve. In practice, this type of function does

TABLE 3.2

Input of Dairy Sector	*Output of Dairy Sector*
0	5
10	8
20	14
30	23
40	35
50	55

not hold good over a long range but in the initial stage of production, it may happen to be so.

3. Diminishing Return Function in Dairying

In this case, every successive increase in the input use in dairy farming results in lesser increase in the output of dairy product. This is more practicable and holds over a long range of production. This function can be derived from Table 3.3.

Obviously, marginal productivity of dairy farming (MPP) goes on declining. In rational farm one can say.

$$\frac{\Delta_1 y}{\Delta_1 x} > \frac{\Delta_2 y}{\Delta_2 y} > \frac{\Delta_3 y}{\Delta_3 y} > \ldots \frac{\Delta_n y}{\Delta_n y}$$

This function is also called second degree polynomial function.

ELASTICITY OF PRODUCTION IN DAIRYING

The elasticity of production (e_p) in dairy as compared to the percentage increase in output in all linear polynomial function, the elasticity of dairy, production can be worked out as

$$e_P = \frac{\Delta y / y}{\Delta x / x} \text{ or } \left(\frac{\Delta y}{y}\right).\left(\frac{x}{\Delta x}\right) \text{or} \left(\frac{x}{y}\right).\left(\frac{\Delta y}{\Delta y}\right)$$

TABLE 3.3

Input in Dairy Sector	*Output in Dairy Sector*
0	5
10	15
20	22
30	27
40	30
50	38

In case of elasticity of production of dairy product following points must be known :

(i) A production function in dairying with an elasticity 1.0 throughout indicate constant return to scale.
(ii) When the elasticity of dairy product is 1.0 MP and AP are equal 1.0 MP = AP.
(iii) A production function for which the elasticity is less than 1.0 throughout refers diminishing returns.
(iv) The elasticity is more than 1.0 upto maximum average product and less than 1.0 between the maximum average and maximum total product of dairying.
(v) It becomes less than zero when the total product of dairying declines.[21]

RATIONAL AND IRRATIONAL STAGES OF PRODUCTION FUNCTION IN DAIRY SECTOR

The input-output in dairy farming can be divided into three stages :

(a) Stage-I of Increasing Return
(b) Stage-II of Constant Return
(c) Stage-III of Diminishing Return

Stage-I in dairy sector is characterised by an increasing

output per unit of input. When the supply of variable factor in dairy farming is increased in the initial stages, we get increasing returns because fixed factors are made to work to their full capacity. This increasing returns is attributed to the economies of scale and specialisation. In this stage the average dairy product is increasing. Hence, the marginal product of dairy must be greater than the average product. To maximise the profit an owner or manager of dairy farming can continue to increase variable factor as long as the average productivity is increasing. Stage-I is rational in maximising profit of dairy product. Producer may go at least to the point of highest average dairy products in the use of input.

Stage-II refers to the constant return in dairy farming in which producer can't maintain dairy production in stage-I since they do not have enough resources to extend production into stage-II.

In stage-III the dairy product is declining since the marginal product or the amount of product added by additional units of input is negative. The situation is not profitable to operate with a combination of resources existing in this region. Here the producer is incurring greater cost, as he is utilising more of the variable inputs but getting less output.

A producer whose rational aim is to maximise profit would find two of the above stages, I and III as rational.

Thus, it can be said that for profit maximisation stage-I is the only stage where maximum level of resource use occurs. Regardless of factor cost and product prices, the chosen level of input application should be somewhere in the range between maximum APP (Average Physical Productivity) and this range falls in stage-II.

In this region, the total output of dairy is increasing the marginal products in dairy farming is decreasing and is positive and less than the average dairy product and the average dairy product in decreasing. The small dairy producer who has limited funds may operate in a range of increasing returns to capital and they are not irrational producers. They can maintain production in stage-I since they do not have enough resources to extend production into stage-II.

This study of irrational and rational zone, i.e. three stages of production function shows that dairy production will pay the dairy producer to produce somewhere within stage-II. Hence, in order to determine at what level in stage-II a dairy producer should operate to maximise net revenue, he must have information regarding the price of dairy product and the price of the input.

Once the physical dairy production has been derived, the amount of revenue from particular dairy production process can be determined by multiplying the quantity of dairy product produced by the price of the dairy product. And in this way, the production function has been converted into a revenue function.

For this, if P_y, is the price of dairy product the total revenue (TR) received from particular level can be derived. Similarly, by multiplying the marginal products by the price of the dairy product, we can determine the amount by which the total revenue changes as inputs are added. It is known as value of the marginal product (VMP). In the same way, by multiplying the average product by the price of the dairy product, the value of average product (VAP) can be determined for particular level of input. Thus, VAP (value of average product) is the value of the dairy producer per unit of input at any particular level of input.

If a dairy producer/farmer is familiar with the price of dairy input, and the value of the marginal product (VMP) he can determine the most profitable level of input. If a dairy producer buys input at a given price Px, he will continue to add additional units of the inputs until the value of the marginal product is just equal to the price of the input. Here, it is assumed that the cost of applying the input is included in the price of input and that the price of the input does not change as additional units are applied. At this point, the additional cost of an input is equal to an additional revenue which the dairy input yields. Beyond this level, if more dairy inputs are added, the added cost will be greater than the added return. On the contrary, if fewer units of dairy inputs are used, the value of the marginal dairy product of the input will be greater than the price of input. This situation refers that the return which the producer gets from the additional unit of input is greater than the cost. Thus, situation is more clear.

It is obvious that a dairy producer will get the maximum revenue at a level of input where the cost of marginal unit (Px_1) of dairy product is equal to the value of the additional dairy product (VMP) which he produces. This condition is satisfied at the x level of dairy units. Beyond this level, X_2, VMP is below the price which refers that the added cost is greater the added revenue. Before this level x, VMP is greater than in dairy price which indicates that the marginal product of dairy a higher than the added cost and as such, the dairy producer can increase the level of dairy inputs upto x.

Here the basic condition for maximisation of net revenue is that where the price of marginal product of dairy is equal to the price of dairy inputs. It can be expressed as :

$$P_{y.} = \left(\frac{\Delta_{y.}}{\Delta_{y.}}\right) = P_{x.}$$

This refers that the added return from the last unit should equal to the cost of adding that unit. We can also express the same conditions as :

$$\Delta Py.\ \Delta y. = (Px.\ \Delta x.)$$

Where Py., Δy. is the added revenue and Px., Δx is the added cost of dairy farming. Another way of expressing the condition of the maximum revenue is that the ratio of the price of the input to the price of the product shall be equal to the marginal product, of the input of dairy farming. This can be stated as below :

$$Px./Py = \Delta x./\Delta y$$

All three conditions of maximising revenue and related with the application of a simple variable x to y, while the level of other inputs are considered unchanged, there are three major factors which affect the most profitable level of an input, i.e. :

(i) Price of dairy output,

(ii) Price of dairy input,

(iii) Physical production relationship.

as it affects the marginal dairy product ($\Delta y_1/\Delta x_1$). Input and output prices of dairy change frequently. Since the most profitable level of an input to use in the production of dairy depends upon the price of dairy product as well as the price of dairy input, changes in prices of dairy product of either of these affect the most profitable level of the input of dairy used.

SUPPLY AND DEMAND BEHAVIOUR IN DAIRY FARMING/WHITE REVOLUTION

In dairy products, the behaviour of supply and demand is typical. Due to seasonal nature of dairy production and high perishability of dairy products, the supply also depicts seasonal characteristics. Further, the response of dairy producers to change to economic parameters is quite slow due to a variety of reasons. The demand on the other hand, is regular throughout the year because most of the dairy products are the basic requirements of life. Thus, the elasticity of demand for dairy production is also low.[22]

Regularly the supply and demand of dairy commodities and thus, stabilizing prices of great importance for the producers as well as consumers. It has assumed and WTO added significance in the post-GATT period of globalization due the following reasons :

(i) Any change in the world market has to be viewed to terms of possibilities of export of tradable surplus and import the deficient goods.

(ii) The change in the policies of subsidies in dairy sector has to be made on the basis of impact on dairy production. So their repercussions on dairy farm economy and the overall economy need to be studied.

(iii) Guidance to the further research in dairy farming has to be given by taking into consideration the return to research not only within the economy but also at global level.

(iv) The possibilities of shift in production patterns and guidance to the producers in this regard has to be given on the basis of demand and supply.

ROLE OF INDIGENOUS DAIRY PRODUCTS

Indigenous dairy products have played an important role in the socio-economic life of our people since the time immemorial. As high as 45-50% of total milk produced in India is converted into these products. On the other hand, only 5-6% of milk is converted into the western type of products produced in the organised dairy sector. In spite of such a great importance of the traditional products in our country the hygienic quality and shelf life of these products is very poor due to their manufacture in the unorganised sector mostly by halwai.[23]

CHANGING MARKET PROFILE OF DAIRY INDUSTRY

Milk production has increased about 615 million tonnes in 2005 from 534 million tonnes during 1993-95. Today in the end of 2005 milk output of development as well as developed region is almost as per at about 250 million tonnes as reflected in a study of the FAO Commodities and Trade Division. For example, the Indian sub-continent contributes a substantial 90 percent of the world buffalo milk production overall, in developing world, the milk from buffaloes, goats, sheep and camels accounts for 15 percent of the world milk production.

Dairy market in the new millennium are being increasingly shaped by the win standards of globalisation and liberalisation. The share of this region is expected to be 79% in 2005 increase in milk production. Between 1995 to 2005 total growth percentage is 17% in developed countries, 79% in developing countries and 4% in transitional countries and average growth rate is 5.9%, 38.3% and 2.7% respectively.

In developing countries following factors are moulding the dairy market profile in the developing countries :

(a) Increased urbanisation and income growth,
(b) Health and pleasure, and
(c) Eating out.

Firstly, Increased urbanisation has became a major force in developing world. According to UN projection, 22 MT of 26 urban agglomerations of over 10 million population in 2015 would be in developing countries. It has effected in many ways.

Secondly, the consumer today demands health benefit in what he eats. Dairy products . . . among the top on this basis. They enjoy a positive image in terms of quality and health benefits. Their products are gaining ground.

Thirdly, in many countries the main growth in food expenditure is in the area of food eastern outside the home. Rapid growth of fast food is highly visible manifestation of eating out.

GLOBAL MARKETS OF INDIAN DAIRY

Indian dairy products particularly Mithai's (sweets) have good commercial market in developed countries where the share of food in the overall household expenditure is small. Thus, people in these countries try out exotic foods from other parts of the world. The waves of migration into America have brought people of various cultures together. With *Chinese food* became common in North America. The *Italians* have made *pizza* a global phenomenon. Recent years have generated a wave of *Mexican food* in the USA. The wide popularity of south restaurants in the UK has given a boost to *Indian curries*. Indian Milk specialities have expanded a new wave as . . . markets.[24]

However, to make an industry presence in the world food market, production of ethics products has to be as per international standards. Efforts have already begun in this direction. For this Export Inspection Council of India (EIC) has been set-up by Government of Indias in New Delhi with sub-office at Chennai, Kolkata, Kochi and Mumbai.

The demand of milk peaks during winter months that concides with India's major festivals like *Dushahara, Diwali* as well as weddings. These are celebrated with sweets. Products such as *rasogullas, gulabjamun, rasmalai, lassi, misti dahi,* etc. are usually sold a tins/plastic pots.

A dairy plant set-up in India has to b registered under Milk and Milk products order (MMPO) of Government of India. To achieve viability in the production of traditional milk products, it should handly about 20,000 litres of milk daily.

The daily plant can also conduct its own R and D to new technologies emanating from National Dairy Research Institution (NDRI), National Dairy Development Board (NDDB), Central Food Technological Research Institute (CFTRI), dairy science colleges and departments of state agricultural universities. For raw materials the plant may tap the informal national grid for moving milk solids, operating in the country.

Chhena from Bihar and Bengal move to Kolkata on regular basis, *Khoa* from Indore finds ready market in Mumbai whole from U.P. finds its way to New Delhi, *Paneer* from U.P., Haryana and Gujarat marketed to Delhi, *Peda* of Gujarat (Rajkot) and U.P. (Mathura) is more popular in India and abroad. *Ghee* is more popular. Its mandies are Jodhpur, Hathras, Kurga, Porbandar, Guntur and Erode.

However, these products offer several technological benefits, viz. their mass appeal, lower cost of production, simple manufacturing technique, use of low grade energy sources and requirement of low infrastructure and operational overhead cost as well as upliftment of the rural people through creation of mass employment and also improvement in economic and nutritional status.

The Indian dairy industry has undergone globalisation as a result of formation of World Trade Organisation (WTO) and implementation of the WTO agreement since January 1995.

India has also emerged as the highest milk producing nation in the world. Her higher growth rate of milk production now compares with that of major dairy products exporting nations. India is going to emerge soon as a major milk products exporting nation.

At this juncture, we will be facing a great challenge of stiff competition, particularly from the western countries in the global international market. For overcoming this challenge particularly in Far East and Middle East countries we must be in a position to offer novel dairy products, which are different from the western dairy products and are hygienic with longer shelf life.

Our indigenous dairy products including the sweetmeats are capable of becoming such novelty products provided we are able to improve the hygienic and keeping quality of these products.

CLASSIFICATION OF TRADITIONAL DAIRY PRODUCTS

In India traditional dairy products can be classified into the following broad categories : (*Indian Dairyman*, 1992).

(i) *Fat-rich products*: (Ghee, Makhan and Malai).
(ii) *Concentrated products*: (Khoa, Khoa-based sweets, Rabri and Basundi).
(iii) *Coagulated products*: (Chhena, Chhenna-based sweets and Paneer).
(iv) *Fermented Products*: (Dahi, Misti Dahi, Chakka, Shrikhand and Lassi).
(v) *Frozen products*: (Kulfi and Kulfa).

I. Fat-Rich Products

Buffalo milk is better suited for the manufacture of fat-rich dairy products as compared to cow milk due to its higher fat, bigger size of the globules and higher proportion of solid fat leading to the higher yield, lesser loss of fat in buttermilk or skin milk easier separation of cream or butter and better texture. (J.S. Sidhu, 1996)

Ghee

Ghee is the most important traditional dairy products of India much more work has been done on its physio-chemical aspects than any other products. Various authors have discussed there aspects of Ghee. (R.S. Sherma, 1981).

Chemically Ghee is a complex liquid of glycer ideas (98% triglycerides, 1-2% drghycerideas and 0.2-0-4% monoghycerides) beside, 38-42.5 (mg.)% of phospholiprds, 275-330 mg.% of cholesterol, 0.22-0.33% free fatty acids, 7.2-8.64 mm/g. carbonyls, 298-488 mg.% unsaponifiable matter and a small proportion of fat soluble vitamins A, D, E and K.

The fatty acid composition of buffaloes *ghee* is different from cow ghee. Buffalo ghee has higher melting point, density, specific gravity and sanponification value but lower BR reading, refractive index.

The buffalo *ghee* is whitish with greenish tinger due to the presence of tetrapyrazole pigments—on the other hand, cow *Ghee* is golden yellow in colour due to presence of carotenoids in it.

Ghee is greatly valued in our country for its characteristic flavour. *Ghee* residue that is rich in flavour components and it has been found to enhance the flavour in flat products like butter oil.

The branded *ghee* from organised dairies is in demand because of its generated quality.[25]

Bikaner in Rajasthan has emerged as India's biggest centre for the production of *resagollas*, exceeding the output of Kolkata. It has modern manufacturing facilities.

Today, the organised sector has found new growth of avenue in the production of *mithais*.

A global market economy has also facilitated a free flow of technologies to produce Indian dairy produts. Scraped surface heat exchangers have been used to pasteurised and process Shrikhand. Meet ball portioning machines and doughtnut fryers have been used in the manufacture of *gulabjamun*. *Japanese pastry making* machines are eployed for preparation of *Burfi*, *Shrikhand* has also been produced as an industrial scale.

More and more dairy plants have started production of *mithais* on commercial scale but their impact is limited.

II. Concentrated Products

Khoa

Khoa an important milk product is prepared by continuous boiling of milk until desired concentration of solids (65-72%).

Khoa is a heat denatured and concentrated milk product used for the preparation of khoa-based sweets, like *Gulabjamun*, *Burfi* and *Pedda*, etc. *Khoa* contains 20-25% moisture, 25-37% fat, 17-20% protein, 22.25% lectose, 3.6-3.8% ash and 100-103 ppm of cron depending on whether it is prepared from cow, buffalo or mixed milk.[26]

The quality of khoa is better when made from buffalo milk as *khoa* from cow milk is inferior due to its moist surface, sticky and spandy texture, which is not considered switable for preparation of sweetmeats. The higher emulsifying capacity of buffalo milk fat due to the presence of larger proportion of bytyric acid containing triglycerides (50%) compared to only 37% in cow milk fat may be responsible for smooth and mellowy texture of the khoa.[27]

Good quality of *khoa* can be prepared from cow milk with incorporation of whey protein concentrate 5%. Inspite of a very high concentration of lactose it does to crystalize out due to high viscosity of product as a result of higher solids and congulation of proteins.

It has been observed that *khoa* prepared in iron pan has minimum while that prepared in stainless steel pan has maximum self life. The absolute shelf life of *khoa* can be enhanced to 15-20 days by eliminating the containation with chelating rates.

Khoa prepared from the neutralization of excessive soured milk has a salty tasix. Thus, buffalo milk is preferred to cow milk in making *khoa*. *Khoa* is manufactured primarily by Halwai's in jacketed kettles.[28]

III. Heat and Acid Coagulated Products

Paneer and Chhana are the two important heat and acid coagulated products. A lot of work has been done on the physico-chemical attributes of these products as discussed below:

(a) Paneer

Paneer, the indigenous variety of cheese is obtained by the acid coagulated of heated milk. The phenomenon of coagulation involves the formation of large structured aggregates of proteins in which fat and other colloidar and dissolved solids are entertained with whey. It has been revealed through ultramicroscopy that paneer has granular structure consisting of protein particles having a core and living ultra structure irrespective of type of milk used and is similar to any other milk product obtained by coagulating hot milk acid to a pH in the vicinity of 5.5.[29]

Cooking of fried paneer by boiling in salt water (1.5%) for 5 min. increases the moisture content of the product beyond the initial level and results in partial restructure of the overall structure. The restoration is more so in cow milk *paneer*.

Good quality *paneer* is characterised by a white colour, sweetish, mildly acidic, nutty flavour, body and closely knit texture. Buffalo milk *paneer* is of good quality as it has all there attributed. On the other hand, cow milk *paneer* is of inferior quality as it is too soft, fragile and its pieces lose their identity on cooking.[30]

Protein and calcium tend to contribute firmness and chewiness to *Paneer*. Processing makes paneer soft and spongy and reduces the textural differences in raw *paneer* from cow and buffalo and low fat buffalo milk.

More decrease in acidity in cow milk *paneer* may be due to retention of higher lactose in cow milk *paneer*, which ferment to different acids. The dissolved proportion of calcium, magnesium, phosphorus, sodium, potassium, copper, iron and zinc also increased during storage.

When acidity of this milk does not reduce with the addition of fresh skim milk to 0.20-0.23 the resultant product is of poor quality and the yield is less.

The positive milk with normal acidity is also not suitable for *paneer* making due to week body and texture.

(b) Chhana

A base product for a large variety of Indian delicacies namely—*Rasogulla, Chhana, Gaja, Sandesh, Chamcham, Rasmalai, Pantoha, Rajbhog, Chhana murki* etc. is a heat and acid coagulated product. *Chhana* differs from paneer as no pressure is applied to drain the whey and its pH is slightly higher.[31]

Cow milk is better suited for *chhana* making because it produces *chhana* with soft body and smooth texture which is better for making sweets.

Different treatment like addition of sodium citrate and sodium phosphate, dilution of milk with water, chomogenization of milk, addition of cow milk or skim milk, alteration in coagulation temperature and pH or method of straining have been applied to produce a good quality of *chhana* from buffalo milk.

According to De and Ray, cow milk *chhana* is soft bodied with small grainy smooth texture and more cohesive compared to hard bodied granular hard texture *chhana* from buffalo milk.[32]

Chemical quality of *chhana* is largely dependent upon the initial composition of milk, percentage of moisture retained in the product and losses of milk solids in the whey. Buffalo milk *chhana* retains less moisture and lactose (53.8 and 2.57%) but higher total solids (46.31%) protein (17.10%) and ash (2.03%) compared to 56.57%, 30.6%, 43.43%, 22.37% and 1.5% respectively in cow milk *channa*. Retention of minerals is higher in buffalo milk *chhana*, copper and iron contents are higher in market *chhana* compared to laboratory made *chhana*. Both these products are unsuitable for sweet making.

It is observed that acidic milk with 0.25-0.28% acidity can be utilized with addition 20.2% sodium citrate to milk followed by through washing of curd after coagulation of milk.

Buffalo milk *chhana* is generally harder, chewier, less smooth and less stricky compared with cow milk *chhana*, mixed milk *chhana* is intermediate of the two.

IV. Fermented Products

Chakka, Dahi, Shrikhand, Misti Dahi and *Lassi* are the main products manufactured and consumed in India.

(a) Chakka and Shrikhand

The semisolid curd mass obtained by the removal of whey from curd (Dahi) is commonly known as chakka, which is the base material for making Shrikhand Chakka has milky white colour, smooth texture and mild acidic flavour. The chemical compound of Chakka and Shrikhand (R.S. Patel, 1993) is in cow milk. T.S. 30.41% and 59.82%, fat 13.12% and 8.75%, protein 13.18% and 5.93%, Lactose 3.10% and 47.73% and ash 0.72% and 0.39% and in buffalo milk T.S. 31.65% and 61.06%, fat 19.93% and 6.31%, protein 14.04% and 6.96%, Lactose 3.17% and 47.24% and ash 0.89% and 0.52%.

During 4 days storage body and texture is adversely affected due to hard, coarse and dry surface in chakka.

Shrikhand with distinct taste richness, delicacy diversity and fairly longer shelf life is very popular in India particularly in states like Gujarat and Maharashtra. Reducing the amount of added source can increase protein content of Shrikhand. Lectose content of the product depends on the extent of its degradation, moisture content of chakka and Shrikhand conditions.[33]

Beside fresh milk, other products like diluted condensed milk, reconstituted skin milk, butter milk, skin milk and condensed milk have been used for preparation of Shrikhand. However, quality of Shrikhand is obtained from these product is inferior. Flavour of Shrikhand is influenced by the developed acidity in curd and chakka prior to their convention to Shrikhand.[34]

Buffalo milk is preferred for making Shrikhan due to higher yield and better quality of the finished products obtained from it. Buffalo milk Shrikhand is rich in minerals including calcium, magnesium, phosphorus, copper, iron and zinc compared to cow milk Shrikhand. However, sodium, potassium and chloride are less in buffalo milk shrikhand. The increase in the penetration value during storage is more in cow milk Shrikhand than buffalo milk Shrikhand.

(b) Dahi

Dahi is an indigenous dairy product obtained by the lactic acid fermentation of milk. A good quality of Dahi is of firm and smooth consistency with a sweet aroma and pleasant taste. The composition of Dahi is similar to the milk from which it is produced. Dahi from buffalo milk is superior in body and texture than cow milk Dahi.[35]

Mr. Khambetta and Dstur determined the nutritive value of Dahi and the physico chemicals changes which take place in milk during fermentation. They found that nutritive value of Dahi is less than the milk from which it is produced.[36]

Milk with 4% far produces the curd with highest hardness and further increase in fat had no favourable effect on the texture of curd. Increase in case of Dahi had a favourable effect while the increase in whey protein had an adverse effect on the texture of curd. Homogenization of milk improves the flavour intensity of Dahi.[37]

(c) Misti Dahi

In the eastern region of India sweet variety of Dahi known as *Misti Dahi, Lal Dahi* or *Payondhi* is popular. It is produced with the addition of 6-6.5% sugar of milk either during boiling or at setting stage. Prolong heating of sweetened milk at low temperature leads to milk solid concentration and development of brown colour. The ultrastructure of Misti Dahi shows a clear protein matrix with uniform distribution of lactic acid bacteria and yeast cell.[38]

The market quality of Misti Dahi varies widely due to difference in the quality of raw milk, milk solid concentration, type of culture used and the inoculation temperature.

Increase in the concentration of cane sugar had an adverse effect on the flavour score and curd tension. The optimum concentration of sugar was 1.4%.

(d) Lassi

Lassi is refreshing beverage prepared by stirring up Dahi and adding sugar and small quantity of cold water. *Lassi* is whitish, viseous, rich aroma and mild to highly acidic product.[39]

ADVANTAGE OF KHOA

Khoa is manufactured primarily by *halwais* in Jacketed kettles, which inherently suffer from several disadvantages as given below :

- Wide variations in chemical, microbial and sensory qualities from batch to batch.
- Small scale batch process unsuitable for commercial adoption.
- Low heat transfer coefficients causing equipment to be bulky.
- Excessive strain and fatigue on the operator.
- Poor package.
- Limited shelf life of the product.

Several R/D efforts have been made in the recent past to overcome the above mentioned problems. The first semi-continuous khoa-making machine was developed by Banerjee *et. al.* (1968) and its subsequent modification was done by De and Singh (1970).

Christic and Shah (1992) have designed and developed a three stage unit for continuous khoa manufacture. Khoa has a limited shelf life of about 5 days at 30ºC.

Khoa Powder

Under the sub-tropical conditions prevailing in India the shelf life of khoa can be extended only to a limited period. The prevailing methods cannot help preserving khoa produced in flush season, till the on set of lean months, i.e. for about 6 months manufacture of *khoa powder* has therefore, tremendous scope. Several attempts have been made in this direction. Patel and De (1977) for the first time tried to prepare khoa powder. They prepared khoa by traditional method, made for slumy with water along with the emulsifiers or stabilizers in order the standardized the milk solids to 18% and finally dried the mix separately on drum and spray driers. (Bylund, G., 1995, Dairy Processing Handbood—T.P. Processing System, Sweden, pp. 75-106).

The technologies for manufacture of khoa powder for small and industrial methods have been developed using treys drying and spray/roller drying respectively.

Method A : Khoa → Heating → Adding of sugar @ 30% of Khoa → (Additives → kneading 58º) → Cooling (25ºC) → Forming → Packaging.

Method B : Milk → Standardisation (SNF : Fat 1.5:1) → continuous khoa unit → khoa → (sugar and additives → stephan processing unit) → Burfi Polystyrene tubs setting vacuum packaging → storage.

Burfi

Burfi is a popular milk-based confection in which the base material is essentially khoa. Sugar is added is different proportions and other ingredient incorporated according to the demand of consumers. *Several varieties* of Burfis are sold in the market depending on the additives present, viz. plain Mewa, Pista, nut, chocolate and Rava Burfi. A lot of variation can be observed in physical attributes of market samples. Good quality of Burfi, however, is characterised by moderately sweet, taste, soft and slightly greasy body and smooth texture with very fine grains. Colour, unless it is chocolate Burfi, should be white or slightly yellowish.

The chemical composition of Burfi depends on the quality and composition of raw-materials, amount of sugar and other ingredient and extent of heating. (Chandan, R.C., 1995, other Fermenated Dairy Products, pp. 404-10).

With a view to extend the shelf life and improve the texture of Burfi, 50% replacement of cane sugar with corn syrup (42 DE) has shown desired results. It also reduced the after activity, thereby exerting an inhibitory influence on the growth of the bacteria. The shelf life of Burfi at 30°C in parchment paper is about 10 day. Packaging of Burfi in presterilized (0.5%, H_2O_2) cryovac pouches increased the shelf-life upto 30 days. Filling of Burfi into polystyrene tubs and vacuum packaging increased the shelf-life to more than 60 days.[40]

Peda

Peda or *Doodh Peada* is limited to the use of khoa alone as a base material. As compared with Burfi, it is granular in texture having dry body because of comparatively lower moisture content. There are numerous varieties of Peda. Their methods of manufacture vary from region to region depending upon the consumer's requirements. Consequently, the sensory and chemical attributes of peda vary to a great extent.

Peda is prepared through two methods : (i) Traditional methods, and (ii) Industrial method.

(i) Industrial Methods of Peda

It differs slightly depending on type of *pedas*. Basically, the method is identical to that of Burfi preparation wherein a mixture of khoa and sugar is heated at low fire till desired texture is attained. Peda is made into round balks of about 20-25 g. size, normally by rolling between the palms. The product may also be formed into different shapes and sizes using different dies/moulds. Peda is generally packed in paper board/boxes having a parchment paper liner or grease proof paper liner (Reddy, 1985).

In another approach 168 kg. of peda can be prepared from 600 liters of buffalo milk in a shift of 8th on a set of six crude oil combustion furnaces. In this case 5 liter of milk is taken each time in a pan and when milk comes to boiling, 450 g. sugar is added and subsequently peda is prepared as described above. This method is adopted in rural milk centres Kutch district of Gujarat.[41]

In Mathura district of U.P., khoa is first cooked to brown colour in Ghese and then Peda is prepared from it by blending sugar and other activities.

(ii) Industrial Method of Peda

In sugar dairy Baroda Kesar Peda is prepared by adopting a large scale mechanized process.

> Milk → Continuous Khoa—making Maching → Khoa → Heating to 60°C → Adding of sugar @ 30% of khoa → Flavour and other ingredients → Planetary Mixer → Mixture stored at 5°C for 10n → Farming and Packaging Machine and Peda.

Shelf-life of Peda

The shelf-life of Peda under normal packing conditions at room temperature is about 2 weeks. It can be increased to about 40 days by packaging in pre-sterilized shrink wrap. Replacement of 50% cane sugar with corn syrup reduces the water activity to about 0.6 and enhances the shelf life up to 45 days in addition to improving body and texture.[42]

Gulabjamun

Dhap khoa having 40-45% moisture is normally used for its preparation. Like other sweets, the manufacture of Gulabjamun is also largely in the hands of halwais who adopt small scale batch method though there is large variation in the sensory quality of gulabjamun, the most liked product should have grown colour, smooth and spherical shape, shoft and slightly spongy body free from both lumps and hard central core, uniform granular texture, mildly cooked and oily flavour, free from doughly feel and fully succulent with sugar syrup. It shall have optimum sweetness.[43]

The gross chemical composition of Gulabjamun varies widely depending on numerous factors, such as composition and quality of khoa, preparation of ingredients, sugar syrup concentration, etc. The composition of Gulabjamun, on the drained weight basis, varies in the following range, moisture 25-35%, fat 8.5 to 10.5%, proteins 6-7.6%, ash 0.9-1.0% and total carbohydrates 43.0-48.0%.

The method of Gulabjamun making from Dhap khoa has been standardized. It involves proper blending of 750 gm. khoa, 250 gm maida and 5 gm. backing powder to a homogeneous and smooth dough. Small amount of water can be added in case dough is very hard and does not roll into smooth balls. The mix should be prepared fresh every time. The small balls farmed from dough are deep fried in ghee to golden brown colour and subsequently transferred to 60% sugar syrup maintain at about 60°C. It takes about 2 hours for the balls to completely absorb the sugar syrup.[44]

> Khoa → (Maida + other ingredient → Planetary mixed → Dough → Portioning machine (8 mg. each portion) → Ball farming machine → Deep fat frying tank (140°C) → Sugar syrup tank → Gulabjamun → Packaging → storage.

The shelf-life of Gulabjamun in sugar syrup is 5-7 days which can be extended to 3 weeks by hot filling in polystyrene cups and adding 0.1% potassium sorbate as a preservative.

Gulabjamun Mix Powder

The traditional method of Gulabjamun has several limitations such as non-availability of khoa all the year around, variation in its quality and that the resultant sweet and poor shelf time. Consequently, many Gulabjamun mixes sold in the market are governed by the patient regulation and the general public are not aware about the composition and constituents of these mixes. The Ghosh *et. al.* (1986) have developed the formulation of Gulabjamun mix powder from both roller as well as spray dried skin milk. Sweets of highly uniform and acceptable quality can be prepared by both housewives and confectioners from these mixes. The shelf-life of these mixes in metalized laminated pouches are about 9 months at room temperature.

Rabri

Rabri is a partially concentrated and sweetened milk product containing several layers of clotted cream (*malai*). It is quite popular in northern and eastern parts of the country. Traditionally it is prepared by milk at a very small scale by simmering whole milk for a prolonged period and adding sugar after achieving the desired concentration.

Rabri is generally manufactured and stored in open and shallow type of container which lead to enormous contamination from surroundings.[45]

Based on the market survey, the desirably sensory and chemical attributes of Rabri are characterised and a standard method of manufacture for Rabri is developed. It involves standardization of buffalo milk of 6% fat, its simmering in a steam jacketed kettle at 90°c, repeated removal of clotted cream (*malai*) on the colder part of the kettle or to separate container after removing about 100 gm. clotted cream from 1 kg. milk and adding sugar @ 6% of initial milk to the concentrated milk. The clotted cream is finally added to the concentrated sweetened milk.

> Buffalo milk → SSHE → concentration (–2.5 fold) ← Sugar @ 6% of milk → Sweetened condensed milk ← (Shredded chhana/paneer @ 10% of milk) → Rabri → Packaging.

TESTING OF MILK

(A) Sampling of Milk

According to Dr. M.S. Swaminathan the milk fat tends to rise to the top when milk is kept undisturbed for sometime. For obtaining a representative sample of milk for testing, it is therefore, important that the milk is thoroughly mixed. In the case of daily supplies of milk or milk products such as cream, it may be necessary to prepare a composite sample for testing once a week. This is done by drawing an aliquot sample from the bottle to which a few drops of formation, which acts as a preservative, has been added. The sample drawn each day should be in proportion to the quantity of the milk supplied. It is important that the quantity of main added is not excessive otherwise it will be difficult to dissolve the curd formed when the milk is tested for its fat content.[46]

Milk Fat

The fat in milk is generally determined by Gerber method in the dairies 10 ml. of sulphuric acid to measured into butyrometer acid 11 ml. of well mixed allow it to mix with the acid. With the help of automatic pipette 1 ml. of any alcohol is added to the butyrometer. The contents of the butyrometer are mixed after closing the byturometer with the slipper.

The Gerber method for determination of fat in milk generally gives slightly higher fat percentage than the actual fat present a Indian cow milk. With buffalo milk having a much higher content, these differences increase. The Indian standard institution has therefore, suggested the use of a correction factor for application of the fat percentage determined by the above method. However, in actual practice, the co-operative societies or private individuals do not probably realize that the correction is necessary to be made of the percentage as shown in the butyrometer.

As a result, the plant has to undergo a considerable loss due to higher payment for milk on the basis of fat content of milk. (Swaminathan, M.S., 1990)

Adulteration of Milk

The milk is commonly adulterated by the addition of water with or without a soluble material such as sugar, by subtraction of fat or by addition of skim milk. The addition of water has the effect of lowering the proportion of its constituents and also its specific gravity, which should therefore, give an indication of the degree of purely of the milk. The specific gravity of such diluted milk may however, be raised by the addition of sugar or by the addition of other soluble materials, by removal of fail or by addition of skim milk.

Changes in the normal level of milk constituents are brought out by adding milk powder to milk or by removing fat from milk. On account of short supply of milk in the market, adulteration is common.

On account of the wide variation is the chemical composition of cow and buffalo milk, the latter can be easily adulterated with large quantity of water with skim milk and sold as cow milk. A simple test known as 'Hansa Test' has been developed in National Dairy Research Institute, Karnal and select the presence of buffalo milk in cow milk up to 3%. This test has been successfully used in various states in India. Tests are also available for detecting the adulteration of milk with water. Thickening agents such as starch, sugar and also skim milk and milk powder. Besides, it has been detected that in some cases water thickening agents such as urea, ammonium sulphate or glucose are used.[47]

Urea

The test is based on the development of a characteristics blue or bluish green colour on the addition of 1 ml. of 2% sodium hydroxide solution followed by 0.5 ml. of 2% sodium hydrochloride solution and 0.5 ml. of 5% phenol solution to 1 ml. of filtrate obtained from the suspected sample of milk containing equal quantity of sodium acetate buffer.

Ammonium Sulphate

The test is similar to that the detection of urea except that there is no need of obtain protein free filtrate from milk. The bluish colour turns into deep blue subsequently in the sample of milk containing ammonium sulphate.

Glucose

The suspected sample of milk is heared with equal quantity of Barfords solution for 3 minutes. The development of the colour on the addition of 1 ml. of phosphomulybdie acid reagent to the cooled solution indicates the presence of glucose in the milk.

Table 3.4 deals with the profile of global milk production since 1994 to 2004-05 containing cow milk, buffalo milk, goat milk, sheep milk and other milks. The trend of cow milk and buffalo milk production has been increasing but goat milk and sheep milk production is to some extent constant.

TABLE 3.4

World Milk Production by Species of Milk Animals

(In million Tonnes)

Year	*Cow Milk*	*Buffalo Milk*	*Goat Milk*	*Sheep Milk*	*Others Milk*	*All Milk*
(1)	*(2)*	*(3)*	*(4)*	*(5)*	*(6)*	*(7)*
1994	461.8	50.5	10.0	7.9	1.3	531.6
1995	465.2	54.5	11.7	8.0	1.3	540.7
1996	464.4	57.2	11.8	8.3	1.3	542.9
1997	469.9	59.6	12.1	8.2	1.3	551.6
1998	476.9	61.9	11.6	8.2	1.3	559.8
1999	483.4	64.4	11.4	8.1	1.3	568.7
2000	490.6	66.6	11.6	8.0	1.3	578.1
2001	495.8	69.0	11.8	8.1	1.3	586.1
2002	505.7	73.6	11.8	8.0	1.3	600.5
2003	512.7	75.0	12.0	8.1	1.3	609.1
2004	514.0	76.4	12.6	7.8	1.3	612.1

Source : *IDF Bulletin*, 394/2004 and *Indian Dairyman*, December 2004, Vol. 56, No. 12.

Table 3.5 shows the milk production during the same period in the world. Particularly Africa, North America, South America, Asia, European Union, Japan, CEEC, CIS, Other West Europe, Oceanic and the World.

TABLE 3.5

Cow Milk Production by World Region

(*In million Tonnes*)

Year	Africa	North America	South America	Asia	EU	CEEC	CIS	Other West Europe	Oceanic	World
1994	4.9	85.0	28.7	56.6	120.1	33.0	68.8	5.9	17.2	461.8
1995	5.3	86.0	31.6	58.4	121.8	32.8	62.4	5.9	18.5	465.2
1996	5.2	85.6	33.1	58.4	121.4	32.2	56.3	5.8	19.6	469.4
1997	5.4	87.0	34.3	59.8	121.5	33.3	53.5	5.8	20.2	469.4
1998	5.5	88.1	36.0	61.7	121.2	33.7	53.0	5.8	20.2	476.9
1999	5.3	91.2	36.7	63.7	122.2	32.8	51.6	5.8	22.4	483.4
2000	5.2	93.7	36.6	65.0	121.2	32.0	50.4	5.7	23.6	490.6
2001	4.8	93.0	36.8	66.0	121.6	31.7	51.6	5.6	23.9	495.8
2002	4.7	95.1	37.0	69.0	121.9	32.2	53.0	5.6	24.5	505.7
2003	4.7	95.3	37.1	71.2	122.1	32.6	53.0	5.6	24.9	521.7
2004	4.7	95.4	37.7	73.0	120.7	33.0	52.5	5.6	25.7	514.0

Source : *IDF Bulletin*, 391/2004 and *Indian Dairyman*, December 2004, Vol. 56, No. 12.

Table 3.6 demonstrates global buffalo milk production in million tonnes during the same period of 1994 to 2004 particularly in India, Pakistan, Egypt, Italy and the whole world.

TABLE 3.6
Global Buffalo Milk Production (In M.T.)

Year	*India*	*Pakistan*	*Egypt*	*Italy*	*World*
1994	32.5	13.2	1.4	0.1	50.5
1995	35.7	14.0	1.4	0.1	54.5
1996	37.0	15.0	1.6	0.1	56.2
1997	38.4	15.4	2.0	0.1	59.6
1998	40.1	15.9	2.0	0.2	61.9
1999	41.9	16.5	2.0	0.2	64.4
2000	43.6	16.9	2.1	0.2	66.6
2001	45.3	17.5	2.1	0.2	69.0
2002	47.2	18.0	2.1	0.2	73.6
2003	49.0	18.5	2.1	0.1	75.0
2004	50.0	18.7	2.4	0.1	76.4

Source : *IDF Bulletin*, 394/2004 and *Indian Dairyman*, December 2004, Vol. 56, No. 12.

Table 3.7 presents the picture of demand for consumption of dairy products in the world including high income region, West Europe, USA and China, Japan and Korea, Australia and New Zealand and low income and middle income countries like Central Europe, Russia, Ukrain, China and Hong Kong, South-East Asia, Latin America, Middle, East and North Africa and Sub-Saharan Africa etc.

Table 3.8 undertakes cost analysis of Khoa and Paneer and Ghee comprising fixed cost, variable cost, marketing cost and total cost per kg. products. Fixed cost consists of depreciation on building, equipment and machinary, furniture, interest on fixed capital, licence fees and taxes whereas variable cost comprises milk-labour, water and electricity, fuel, chemicals and detergents packaging and storage, etc. and the marketing cost contains transportation, labour, etc.

TABLE 3.7
Demand for Dairy Products by Region

Region	*Consumption 2000 (mm MT Milk equiv.)*	*Average Annual change 2001-06 (estimated)*
World	572	2%
High Income Region	, 224	1%
Wester Europe	119	1%
USA and China	85	1%
Japan and Korea Republic	13	1%
Australia and New Zealand	7	+ 0.1–1%
Low and Middle Income Region	335	+ 2–3%
Central Europe	33	1%
Russia and Ukraine	60	1%
China and Hongkong	10	4%
South Asia	110	+ 2–3%
South East Asia	7	4%
Latin America	68	3%
Middle East and North Africa	29	+ 2–3%
Sub-Saharan Africa	18	1%
Others	13	1%

Source : ZMP and Rebobank International 2001 and Agriculture and Industry Survey, Vol. 12, Nos. 9 and 10, 2002, p. 18.

Table 3.9 explains demand for dairy products particularly Khoa, Paneer and Ghee comprising per kg. prices recovered, net cost of production per kg. net profit in Rupees, actual production, rate of return per rupee invested and the profit generated.

Table 3.10 demonstrates per capita milk production and consumption and intake of calories and protein in selected Asian countries in the year 2000 of India, Bangladesh, Pakistan, China, Thailand, Vietnam, Sri Lanka, Myanmar, Indonesia, East and South East Asia, South Africa, Philippines, Laos, Malaysia, etc.

Table 3.8
Cost Analysis of Khoa, Paneer and Ghee

	Khoa		Paneer		Ghee	
	Cost	*Share*	*Cost*	*Share*	*Cost*	*Share*
(1)	*(2)*	*(3)*	*(4)*	*(5)*	*(6)*	*(7)*
A. Fixed Cost						
1. Depreciation on :						
(a) Building @ 10%	0.11	0.17	0.12	0.19	0.20	0.15
(b) Equip. and Machinary @ 15%	0.21	0.33	0.08	0.13	0.22	0.16
(c) Furniture and Fixtures @ 20%	0.04	0.06	0.05	0.08	0.05	0.04
2. Int. Fixed Capital @ 13%	0.66	1.04	0.77	1.23	1.43	1.04
3. Licence Fee and Taxes	0.01	0.02	0.01	0.01	0.01	0.01
Total Fixed Cost	1.03	1.62	1.03	1.64	1.91	1.41
B. Variable Costs						
1. Milk	47.56	74.60	55.00	87.68	124.08	90.65
2. Labour	3.50	5.49	3.36	5.38	4.00	2.92
3. Water and Electricity	0.07	0.11	0.05	0.08	0.09	0.07
4. Fuel	6.30	9.89	1.82	2.90	5.82	4.25

5. Chemicals and Detergents	0.16	0.25	0.55	0.88	0.39	0.27
6. Packaging and Storage	0.32	0.50	0.36	0.57	0.01	0.01
Total Variable costs	57.91	90.84	61.14	97.47	134.39	98.18
C. Marketing Costs						
1. Transporttion	1.03	1.61	—	—	—	—
2. Labour	1.10	1.73	0.56	0.89	0.58	0.42
3. Adat	2.68	4.20	—	—	—	—
Total Marketing cost	4.81	2.54	—	—	—	—
D. Total Cost per kg. product	63.75	100.0	62.7	100.0	136.88	100.0

Source : *Indian Dairyman*, December 2004, Vol. 56, No. 8, p. 75.

Table 3.9
Demand for Dairy Products by Region

Particulars	Khoa	Paneer	Ghee
1. Prices recovered (Rs./kg.)	70.00	70.00	150.00
2. Net Cost of Production (Rs./kg.)	63.75	62.73	136.68
3. Net Profit per kg. in Rupee	6.25	7.27	13.12
4. Breakeven Productions 1 kg./year	2504.53	1641.69	1204.85
5. Actual Production (% of BEP)	952.82	311.26	302.94
6. Rate of return per Rupee invested	1.10	1.11	1.09
7. Profit generated (In 000 Rupees)	149.15	5.11	3.65

Source : *Indian Dairyman*, December 2004, Vol. 56, No. 8, p. 75.

Table 3.10
Per Capita Milk Production and Consumption and Intake of Calories and Protein in Selected Asian Countries in the Year 2000

Countries	*Per Capita Milk Products*	*Per Capita Milk Consumption*	*Per capita Calories intake (kcl per day)*	*Per Capita Protein in per day*
(1)	*(2)*	*(3)*	*(4)*	*(5)*
Bangladesh	15.7	16.7	27	1
Myanmar	13.1	19.6	30	2
Sri Lanka	15.7	41.9	72	4
India	82.4	81.2	158	7
Indonesia	3.8	7.7	12	1
Laos	1.2	4.1	8	0
Malaysia	2.3	54.0	116	4
Nepal	52.0	50.0	102	4
Pakistan	191.1	156.8	341	16
Philippines	0.1	23.1	30	2
Thailand	7.6	21.3	35	2

(*Contd.*)

TABLE 3.10 (Contd.)

(1)	*(2)*	*(3)*	*(4)*	*(5)*
Vietnam	1.1	4.7	7	0
China	9.7	11.1	21	1
South Africa	85.4	82.2	162	7
E. and S.E. Asia	8.2	16.3	26	1
Asia	44.0	48.7	90	4
World	96.7	94.4	145	7

Ref. : Quoted from Taniji Birhas (2004) Calculated by then based on FAO Statistics.

Source : *Indian Dairyman*, February 2004, Vol. 57, No. 2.

SUSTAINABLE DEVELOPMENT OF DAIRY SECTOR/ WHITE REVOLUTION

There are many areas in which research and development efforts made have brought about and can bring about many positive implications to the development of Cottage dairy industries. Effort has been made to shortest the recent development that have helped in reducing the cost of milk production at a farmgate added value to the farmer's produce and his price realization and the organisation innovation that can help network the scattered milk producer. For the cottage dairy industry to succeed it is important that the cost of milk production is low at the farmgate. This is achievable by increasing the per animal milk production increasing lifetime production per animal. (Goodland, Robert, 1977, Environmental Sustainability and Agriculture : Diet Matters, Ecological Economies 23(3) 189, 2000).

Thus, sustainable dairy development can be possible by keeping such breeds of cattle and buffaloes that world required minimal and inexpensive external inputs, are efficient converters of by-products of agricultural crops and of the processed foods, can survive the harsh weather and system in rural environments and are resistant to local and exotic disease.

Dairy development or second phase of white revolution forms the basis of sustainable development in Indian economy because it is based on farmers with small holding and milk production per animal.

The underlying objective of globalisation and trade liberalisation has been to create a single world market. The advantage would be those who produce cheapest product should market it globally.

But most of the developed countries also provide heavy subsidies under three boxes of WTO :

(i) *Green Box*,
(ii) *Blue Box*, and
(iii) *Amber Box* under the provisions of:

(a) *Market Access*,
(b) *Domestic support*, and
(c) *Export subsidies*. (WTO, Hongkong Ministerial Conference, December, 2005).

Thus, under dairy market their dairy products are subsidised heavily for increasing exports of dairy products to maintain sustainable economic growth. However, it has created unhealthy competition in the domestic market of importing countries.

Developing country like India had committed zero percent base and bound rates in imports of skimmed and full cream milk powder and 40% as butterfat, cheese and whey under the WTO agreement.

Developed countries provide huge subsidies to its milk producers but the poor and developing countries can not afford to provide any doles despite farmer's inability to realize era the cost of milk production. [Alagh, Y.K., 2002; Key Note Address Dairy Industry Conference, Mumbai, *Indian Dairyman*, 54(2)]. India's traditional dairy products sector is poised for rapid expansion with the result of application of modern process technologies in the production of indigenous sweets (Mithais). The rising demand for packaged, fresh dairy products like dahi, paneer, lassi and cream in widening the base of the modern dairy sector. (Aeja, R.P., 1997; *'Traditional Milk Specialities'*, Dairy India, Dairy India Yearbook, Delhi, pp. 376).

Significantly, technological innovations have core at a time when ethnic milk products are attracting world-wide attention.

In the new millennium, the global dairy industry is in search of initiatives to enlarge its market. Dairy markets in the new millennium are being increasingly shaped by the twin standards of globalisation.

The traditional dairy products of Indian sub-continent are broadly classified into following five categories : (Acharya, K.J., *Historical Dictionary of Indian Food*).

(a) Desiccated Milk-based Products, i.e. Khoa, Peda, Burfi, Gulabjamun, etc.
(b) Heat Acid coagulated products, i.e. paneer, channa, rasogulla, etc.
(c) Cultured/Fermented products, i.e. Dahi, Lassi, Shrikhand, etc.
(d) Fat—Rich product, and i.e. Ghee, Makhan, Melai, etc.
(e) Milk-based Pudding/Deserts, i.e. Kheer, Kaju Burfi, Burfi, Payasam, Sohan, Halwa, Gajara ka Halwa, etc.

According to the Report of the National Commission of Agriculture, 2002, Vol. VII, Animal Husbandry, Government of India, Dairying is a significant source of generating rural income and employment. It provides supplementary income to the farmer households and utilises the idle family members available in the farmers' families. Table 3.11 shows profitability of various products whereas Table 3.12 highlights milk requirements, raw-material cost and sales realization.

Table 3.13 explains production cost breakdown and not profit for 14 selected milk powder in a very clear manner.

Table 3.14 presents a summary of parameters of various dairy plants in all four regions of the country.

Financial Analysis

In the Table 3.15 to 3.18, various aspects of the economic viability of the mechanized production of Indian Milk viability of

TABLE 3.11

Profitability of Various Products as Gross Percentage of Cost

(per cent)

Milk	*Raw Mtl.*	*Servi-ces*	*Packa-ging*	*Sto-rage and distb.*	*Salary and wages*	*Depre-ciation and interest*	*Total cost*	*Net profit*
Shrikhand	43	6	14	5	7	8	83	17
Dahi	34	8	25	5	6	6	84	16
Mishti Dahi	40	6	22	5	4	3	80	20
Lassi	40	7	7	4	6	6	70	30
Peda	54	5	10	4	6	7	86	14
Burfi	51	5	9	4	6	7	82	18
Gulabjamun	33	5	14	5	3.5	3.5	64	36
Kheer	33	4	17	3	4	4	65	35
Basundi	39	4	14	3	6	6	72	28
Pal Payasam	31	5	21	4	4	4	71	29
Paneer	58	3	4	3	10	11	89	11
Rasogolla	33	5	14	3	5	5	65	35
Sandesh	42	3	7	3	8	8	71	29
Ghee	82	2	3	3	2	2	94	6

TABLE 3.12

Milk Requirement, Raw-material Cost and Sales Realization for 14 Selected Indian Milk Products

Description	*Qty. (kg./day)*	*Rate (Rs./kg)*	*Value (Rs./day)*
(1)	*(2)*	*(3)*	*(4)*
SHRIKHAND (Ratio 2.7)*			
Raw-materials		31.10	94,640
Milk (F–0.1%, SNF–9%)	8,500	6.40	54,400
Cream (F–80%)	210	72.00	15,120
Sugar	1,360	15.00	20,400
Other Ingredients			6,720

(Contd.)

TABLE 3.12 (*Contd.*)

(1)	*(2)*	*(3)*	*(4)*
Net Output and Sales Realization	3,107	72.00	2,23,704
Gross Output	3,270		
Handling Loss (5%)	163		
DAHI (Ratio 1.0)*			
Raw-materials		10.95	53,640
Milk (F–4%, SNF–9%)	5,000	10.00	50,000
Skimmed Milk Powder (SMP)	52	70.00	3,640
Net Output and Sales Realization	4,900	32.00	1,56,800
Gross Output	5,000		
Handling Loss (2%)	100		
MISHTI DAHI (Ratio 0.7)*			
Raw-materials		14.88	72,916
Milk (F–3%, SNF–9%)	3,563	9.00	32,067
SMP	271	70.00	18,970
Sugar	900	15.00	13,500
Cream (F–35%)	266	31.50	8,379
Net Output and Sales Realization	4,900	45.00	2,20,500
Gross Output	5,000		
Handling Loss (2%)	100		
LASSI (Ratio 0.8)*			
Raw-materials		10.02	61,975
Milk (F–3.8%, SNF–9%)	5,000	9.82	49,100
Sugar	750	15.00	11,250
Other Ingredients			1,625
Net Output and Sales Realization	6,184	25.00	1,54,600
Gross Output (litres)	6,310		
Handling Loss (2%)	126		
PEDA (Ratio 3.3)*			
Raw-materials		54.17	83,420
Milk (F–5%, SNF–9%)	5,150	11.80	60,770
Khoa Output (TS 70%)	1,107		

(*Contd.*)

TABLE 3.12 (Contd.)

(1)	(2)	(3)	(4)
Sugar (30% of peda weight)	474	15.00	7,100
Other Ingredients			15,440
Net Output and Sales Realization	1,540	100.00	1,54,000
Gross Output	1,581		
Handling Loss (2.5%)	41		
BURFI/KALAKAND (Ratio 3.6)*			
Raw-materials		56.63	79,850
Milk (F–6%, SNF–9%)	5,150	11.80	60,770
Khoa Output (TS 70%)	1,107		
Sugar (30% of Khoa)	332	15.00	4,980
Other Ingredients			14,100
Net Output and Sales Realization	1,410	110.00	1,55,100
Gross Output	1,440		
Handling Loss (2%)	28.7		
GULABJAMUN (Ratio 1.03)*			
Raw-materials		23.33	20,814
Milk (F–6%, SNF–9%)	900	11.80	10,620
Khoa Output (TS 67%)	200		
Wheat Flour (10%)	20	12.00	240
Baking Powder (2%)	4	91.00	364
Sugar	504	15.00	7,560
Vegetable Fat	58	35.00	2,030
Net Output and Sales Realization	892	70.00	62,440
Gross Output	450		
Handling Loss (1%)	4.5		
Ingredients	120	15.00	1,800
KHEER (Ratio 1.25)*			
Raw-materials		19.68	15,740
Milk (F–5%, SNF–9%)	1,000	10.90	10,900
Sugar	120	15.00	1,800
Ingredients			2,040

(Contd.)

TABLE 3.12 (*Contd.*)

(1)	*(2)*	*(3)*	*(4)*
Net Output and Sales Realization	800	60.00	48,000
Gross Output	808		
Handling Loss (1%)	8		
BASUNDI (Ratio 2.0)*			
Raw-materials		27.35	13,400
Milk (F–5%, SNF–9%)	1,000	10.90	10,900
Sugar	100	15.00	1,500
Ingredients			1,000
Net Output and Sales Realization	490	70.00	34,300
Gross Output	500		
Handling Loss (5%)	10		
PAL PAYASAM (Ratio 1.12)*			
Raw-materials		14.85	13,225
Milk (F–3.5%, SNF–8.5%)	1,000	8.80	8,800
Rice	70	15.00	1,050
Sugar	225	15.00	3,375
Net Output and Sales Realization	890	48.00	42,720
Gross Output	908		
Handling Loss (5%)	18		
PANEER (Ratio 5.0)*			
Raw-materials		51.75	1,03,500
Milk (F–5.5%, SNF–9%)	10,000	10.35	1,03,500
Net Output and Sales Realization	2,000	90.00	1,80,000
RASOGOLLA (Ratio 1.75)*			
Raw-materials		23.07	13,150
Milk (F–4%, SNF–9%)	1,000	9.30	9,300
Sugar	59	15.00	886
Net Output and Sales Realization	570	70.00	39,900
Gross Output	300		
Handling Loss (5%)	15		
Sugar Syrup for Packing	285		

(*Contd.*)

TABLE 3.12 (*Contd.*)

(1)	*(2)*	*(3)*	*(4)*
SANDESH (Ratio 6.2)*			
Raw-materials		62.22	10,186
Milk (F–4%, SNF–8.5%)	1,000	9.30	9,300
Sugar	59	15.00	886
Net Output and Sales Realization	162	150.00	24,300
Gross Output	166		
Handling Loss (2.5%)	4		
GHEE (Ratio 1.3)*			
Raw-materials			
Cream (F–80%)	1,200	76.00	91,200
Net Output and Sales Realization	922	120.00	1,10,640
Gross Output	960		
Handling Loss (5%)	38		

*Ration refers to the volume of milk required to produce one kg. of the product.

Source : Aneja, R.P.; Mathur, B.N.; Banerjee, A.K. (2002) : Technology of Indian Milk Products, A Dairy India Publication, Delhi, India, p. 221.

the mechanized production of Indian milk products have been analyzed and presented in terms of the following parameters:

- Gross Margin (Sales Realization—Raw-material Cost).
- Operating Margin (Sales Realization—Production Cost).
- Total Cost (Production Cost + Fixed Cost)
- Net Profit (Sales Realization—Total Cost)
- Net Profit as percentage of Sales Realization.
- Sales Realization as percentage of Capital Cost.

Analysis of the cost distribution of all products presented in table shows that the production of chhana-based Bengali sweets improves profitability because these products give higher margins.

TABLE 3.13

Production Cost Breakdown and Net Profit for 14 Selected Indian Milk Products

Sl. No.	Items	Dahi Milk: 8,500 kg. (F-0.1%, SNF-9%) Product: 3,107 kg.		Misti Dahi Milk: 5,000 kg. (F-4%, SNF-9%) Product: 4,900 kg.		Lassi Milk: 3,563 kg. (F-3%, SNF-9%) Product: 4,900 kg.		Shrikhand Milk: 5,000 kg. (F-3.8%, SNF-9%) Product: 6,184 kg.	
		Rs./kg.	Total (Rs.)	Rs./kg.	Total (Rs.)	Rs./kg.	Total (Rs.)	Rs./kg.	Total (Rs.)
	(1)	(2)	(3)	(4)	(5)	(6)	(7)	(8)	(9)
I.	Expenditure								
	1. Raw-Material Cost	31.10	96,640	10.95	53,640	14.88	72,916	10.02	61.975
	Milk 6.40	54,400	10.00	50,000	9.00	32,067	9.82	49,100	
	Others		42,240		3,640		40,849		12,875
	2. Operating Cost		71,831		68,045		81,771		34,106
	A. Service Cost		13,826		12,005		12,838		8,657
	A.1 Power (@ Rs. 4.75/kwh)	1.00	3,107	0.35	1,715	0.36	1,764	0.25	1,546
	A.2 Steam (@ Rs. 0.90/ kg. of steam)	0.45	1,398	0.10	490	0.26	1,274	0.15	927
	A.3 Water & Chemicals	1.00	3,107	1.00	4,900	1.00	4,900	0.50	3,092
	A.4 Maintenance & Services	2.00	6,214	1.00	4,900	1.00	4,900	0.50	3,092
	B. Packaging Cost	10.00	31,070	8.00	39,200	10.00	49,000	1.66	10,265

(Contd.)

TABLE 3.13 (*Contd.*)

(1)	(2)	(3)	(4)	(5)	(6)	(7)	(8)	(9)
C. Marketing Cost		11,185		7,840		11,025		6,184
C.1 Sales Promotion	1.43	4,474	0.64	3,136	0.9	4,410	0.25	1,546
C.2 Storage & Distribution	2.15	6,711	0.96	4,704	1.35	6,615	0.75	4,638
D. SALARIES & WAGES	4.92	15,300	1.83	9,000	1.81	8,908	1.45	9,000
3. Production Cost (1+2)	54.14	1,68,221	24.83	1,21,685	34.74	1,54,687	15.93	96,081
4. Fixed Cost: Depreciation and Interest	5.47	17,000	2.04	10,000	1.45	7,126	1.61	10,000
5. Grand Total (3 + 4)		1,85,221		1,31,685		1,61,813		1,06,081
II. Income	72.00	2,23,704	32.00	1,56,800	45.00	2,20,500	25.00	1,54,600
6. Sales Realization								
III. Financial Analysis								
Gross Marging (6-1)	40.83	1,27,064	21.05	1,03,160	26.94	1,47,584	14.98	92,625
Opening Surplus/Marging (6-3)		55,483		35,115		65,813		58,519
Net Profit (6-5)		38,483		25,155		58,587		48,519
Net Profit (Percentage)	17.20		16.01		26.60		31.38	

(*Contd.*)

Table 3.13 (Contd.)

Sl. No.	Items	Peda Milk: 5,150 kg (F-6%, SNF-9%) Product: 1,540 kg.			Burfi/Kalakand Milk: 5,150 kg. (F-6%, SNF-9%) Product: 1,410 kg.			Guiabjamun Milk: 900 kg. (F-6%, SNF-9%) Product: 892 kg.		
		Rs./kg.	Total (Rs.)		Rs./kg.	Total (Rs.)		Rs./kg.	Total (Rs.)	
(1)		(10)	(11)		(12)	(13)		(14)	(15)	
I.	Expenditure									
1.	Raw Material Cost	54.17		83,420	56.63		79,850	23.30		20,814
	Milk	11.80	60,770		11.80	60,770		11.80	10,620	
	Others		22,650			19,080			10,194	
2.	Operating Cost			39,099			37,145			17,183
	A. Service Cost			8,269			7,571			3,300
	A.I Power (@ Rs. 4.75/Kwh)	1.62	2,494		1.62	2,284	0.61		544	
	A.2 Steam (@ Rs. 0.90/Kg of Steam)	12.00	3,080		2.00	2,820	1.34		1,195	
	A.3 Water and Chemicals	0.75	1,155		0.75	1,057	0.75		669	
	A.4 Maintenance and Services	1.00	1,540		1.00	1,410	1.00		892	
	B. Packaging Cost	10.00		15,400	10.00		14,1 00	10.00		8,920
	C. Marketing Cost			6,160			6,204			3,123

(Contd.)

Table 3.13 (Contd.)

(1)	(10)	(11)		(12)	(13)		(14)	(15)	
C.I Sales Promotion	1.00	1,540		1.10	1,551	1.40	1,249		
C.2 Storage & Distribution	3.00	4.620		3.30	4,653	2.10	1,874		
D. Salaries & Wages	6.01		9,270	1.80		9,270	2.0e		1,840
3. Production Cost (1 + 2)	79.55					1,16,995	.		37,997
4. Fixed Cost: Depreciation and								-	
Interest	6.88	1,22,519	7.30		10,300	2.06		1,840	
5. Grand Total (3 + 4)	6.68		10,300			1,27,295			39,837
II. Income	100,00		1,54,000				1,32,819		
6. Sales Realization				110.00		1,55,100	70.0		62,440
III. Financial Analysis									
Gross Marging (6-1)				53.36		75,250	47.41		41,626
Opening Surplus/Marging (6-3)	45.83		70,580			38,105			24,443
Net Profit (6-5)			31,481			27,805			22,603
Net Profit (Percentage)	13.75		21,181	17.93			36.19		

(Contd.)

Table 3.13 (Contd.)

Sl. No.	Items	KHEER Milk: 1,000 kg. (F-5%, SNF-9%) Product: 800 kg.			BASUNDI Milk: 1,000 kg. (F-5%,SNF-9%) Product: 490 kg.			PALPAYASAM Milk: 1,000 kg. (F-3.5%, SNF-9%) Product: 890 kg.		
		Rs./kg.	Total (Rs.)		Rs./kg.	Total (Rs.)		Rs./kg.	Total (Rs.)	
	(1)	(16)	(17)		(18)	(19)		(20)	(21)	
I.	Expenditure									
	1. Raw-Material Cost	19.68		15,740	27.35		13,400	14.85		13,225
	Milk	10.90	10,900		10.90	10,900		8.80	8,800	
	Others		4,840			2,500			4,425	
	2. Operating Cost			13,328			9,447			15,529
	A. Service Cost			1,888			1,518			2,179
	A.I Power (@ Rs. 4.75/kwh.)	0.17	136		0.30	147		0.14	128.16	
	A.2 Steam (@ Rs. 0.90/ kg. of steam)	0.69	552		1.30	637		0.60	712	
	A.3 Water and Chemicals	0.75	600		0.75	367		0.75	667	
	A.4 Maintenance and Services	0.75	600		0.75	367		0.75	667	
	B. Packaging Cost	10.00		8,000	10.00		4,900	10.00		8,900
	C. Marketing Cost			1,440			1,029			2,670

(Contd.)

TABLE 3.13 (Contd.)

(1)	(16)	(17)		(18)	(19)		(20)	(21)	
C.I Sales Promotion	0.60	480		0.70	343		1.00	890	
C.2 Storage and Distribution	1.20	960		1.40	686		2.00	1,780	
D. Salaries and Wages	2.50		2,000	4.08		2,000	2.00		1,780
3. Production Cost (1+ 2)			29,068			22,847			28,754
4. Fixed Cost: Depreciation and Interest	2.50		2,000	4.08		2,000	2.00		1,780
5. Grand Total (3 + 4)			31,068			24,847			30,534
II. Income									
6. Sales Realization	60.00		48,000	70.00		34,300	48.00		42,720
III. Financial Analysis									
Gross Marging (6-1)	40.32		32,260	42.65		20,990	33.14		29,495
Opening Surplus/Marging (6-3)			18,932			11,453			13,966
Net Profit (6-5)		16,932			9,453				12,186
Net Profit (Percentage)	35.27			27.55			28.52		

(Contd.)

TABLE 3.13 (*Contd.*)

Sl. No.	Items Paneer	Rasogolla Milk: 1,000 kg. (F-5.5%, SNF-9%) Product : 2,000 kg.		Sandesh Milk: 1,000 kg. (F-4%, SNF-8.5%) Product: 570 kg.		Ghee Milk: 1,000 kg. (F-4%, SNF-8.5%) Product: 162 kg.		Milk: 1,000 kg. (Fat 80%) Product: 922 kg.	
		Rs./kg.	Total (Rs.)	Rs./kg.	Total (Rs.)	Rs./kg.	Total (Rs.)	Rs./kg.	Total (Rs.)
	(1)	(22)	(23)	(24)	(25)	(26)	(27)	(28)	(29)
I.	Expenditure								
	1. Raw-Material Cost	10.35	1,03,500	23.07	13,150	62.88	10,186	76.00	91,200
	Milk	10.35	1,03,500	9.30	9,300	9.30	9,300		
	Others	—		3,850		886		—	
	2. Operating Cost		36,600		10,686		4,997		10,539
	A. Service Cost		5,200		1,789		648		2,055
	A.1 Power (@ Rs. 4.75/kwh)	0.25	500	0.32	182	0.74	120	0.58	534
	A.2 Steam (@ Rs. 0.90/kg. of steam)	0.60	1,200	0.82	467	1.26	204	0.65	599
	A.3 Water and Chemicals	0.75	1,500	1.00	570	1.00	162	0.50	461
	AA Maintenance and Services	1.00	2,000	1.00	570	1.00	162	0.50	461
	B. Packaging Cost	4.00	8,000	10.00	5,700	10.00	1,620	3.00	2,766
	C. Marketing Cost		5,400		1,197		729		3,318

(*Contd.*)

TABLE 3.13 (*Contd.*)

(1)	*(22)*	*(23)*	*(24)*	*(25)*	*(26)*	*(27)*	*(28)*	*(29)*
C.1 Sales Promotion	0.90	1,800	0.70	399	1.50	243	1.20	1,106
C.2 Storage and Distribution	1.80	3,600	1AO	798	3.00	486	2AO	2,212
D. Salaries and Wages	9.00	18,000	3.50	2,000	12.30	2,000	2.60	2,400
3. Production Cost (1+2)		1,40,100		23,836		15,183		1,01,739
4. Fixed Cost: Depreciation and Interest	10.00	20,000	3.50	2,000	12.30	2,000	2.60	2,400
5. Grand Total (3 + 4)		1,60,100		25,836		17,183		1,04,139
II. Income								
6. Sales Realization	90.00	1,80,000	70.00	39,900	150.00	24,300	120.00	1,10,640
III. Financial Analysis								
Gross Marging (6-1)	38.25	76,500	46.92	26,750	87.12	14,114	21.08	19,440
Opening Surplus/Marging (6-3)		39,900		16,064		9,117		8,901
Net Profit (6-5)		19,900		14,064		7,117		6,501
Net Profit (Percentage)	11.05		35.24		29.28		5.87	

Source : Aneja, R.P.; Mathur, B.N.; Chandan, R.C. & Banerjee, A.K. (2002) : A Dairy India Publication Technology of Indian Milk Products, p. 222.

TABLE 3.14

A Summary of Parameters of the Proposed Dairy Plant in any of India's Four Regions

(per year)

Items	*Northern Region*	*Western Region*	*Southern Region*	*Eastern Region*
Quantity of milk processed (tonnes)	6,060	6,000	6,050	6,070
Quantity of products manufactured (tonnes)	3,300	3,290	4,520	3,790
Total production cost (Rs. Million)	130	128	135	152
Operating surplus (Rs. Million)	47	45	50	46
Net Profit (Rs. Million)	35	33	38	38
Net profit (as percentage of sales)	20	19	23	18
Capital investment (Rs. Million)	54	54	54	54
Annual turnover (Rs. Million)	177	172	166	209
Turnover to Capital Ratio	3.3	3.2	3.1	3.9

Source : Aneja, Chandan, Banerjee (2002) : *op. cit.*, p. 225.

TABLE 3.15

Breakdown of Operating Surlpus and Net Profit Per Year for a Dairy Plant in North India

Milk Product	Quantity Milk Processed (kg/day)	Product mftd. (No. of days)	Quantity Processed/ year ('000 tonnes)	Quantity Production/ year ('000 tonnes)	Sales per year (Rs. million)	Operating Surplus (Rs. million	Net Profit (Rs. million)
Dahi	5,000	200	1.00	0.98	31.36	6.57	4.57
Mishti Dahi	3,563	50	0.18	0.25	11.03	3.16	2.79
Lassi	5,000	120	0.61	0.74	18.52	6.72	5.52
Peda	5,150	125	0.64	0.19	19.20	3.93	2.68
Burfi/Kalakand	5,150	125	0.64	0.18	19.38	4.76	3.60
Gulabjamun	920	200	0.18	0.18	12.50	4.87	4.52
Kheer	1000	50	0.05	0.04	2.40	0.92	0.82
Paneer	10,000	230	2.30	0.46	41.40	9.17	4.57
Rasogolla	2,000	200	0.40	0.23	15.90	6.42	5.63
Ghee (F–80%)	1,200	50	0.06	0.05	5.50	0.44	0.33
Total		300	6.06	3.30	177.19	46.96	35.03

Source : Aneja, Chandan, Banerjee (2002) : *op. cit.*, p. 225.

TABLE 3.16

Breakdown of Operating Surplus and Net Profit Per Year for a Dairy Plant in West India

Milk Product	*Quantity Milk Processed (kg/day)*	*Product mftd. (No. of days)*	*Quantity Processed/ year ('000 tonnes)*	*Quantity Production/ year ('000 tonnes)*	*Sales per year (Rs. million)*	*Operating Surplus (Rs. million*	*Net Profit (Rs. million)*
Shrikhand	8,500	120	1.02	0.37	26.84	6.62	4.56
Dahi	5,000	250	1.25	1.23	39.20	8.08	5.70
Lassi	5,000	120	0.60	0.74	18.55	7.68	6.49
Peda	5,150	100	0.52	0.15	15.40	3.14	2.14
Burfi/Kalakand	5,150	100	0.52	0.14	15.51	3.80	2.88
Gulabjamun	900	200	0.18	0.18	12.50	4.88	4.52
Basundi	1,000	150	0.15	0.08	5.15	1.99	1.69
Paneer	10,000	150	1.50	0.33	29.70	5.98	2.99
Rasogolla	2,000	100	0.20	0.04	3.30	2.10	1.70
Ghee (F–80%)		300	6.00	3.29	171.65	44.71	33.00
Total		300	6.06	3.30	177.19	46.96	35.03

Source : Aneja, Chandan, Banerjee (2002) : *op. cit.*, p. 225.

TABLE 3.17

Breakdown of Operating Surplus and Net Profit Per Year for a Dairy Plant in South India

Milk Product	*Quantity Milk Processed (kg/day)*	*Product mftd. (No. of days)*	*Quantity Processed/ year ('000 tonnes)*	*Quantity Production/ year ('000 tonnes)*	*Sales per year (Rs. million)*	*Operating Surplus (Rs. million*	*Net Profit (Rs. million)*
Shrikhand	8,500	50	0.43	0.16	11.15	2.75	1.90
Dahi	7,500	250	1.83	1.83	58.80	12.32	8.57
Lassi	5,000	250	1.25	1.54	38.65	16.04	13.54
Peda	5,150	100	0.52	0.15	15.40	3.14	2.14
Burfi/Kalakand	5,150	100	0.52	0.14	15.51	3.80	2.88
Gulabjamun	900	150	0.14	0.13	9.38	3.66	3.39
Pal Payasam	1,000	200	0.20	0.18	8.55	2.79	2.43
Paneer	10,000	70	0.70	0.19	0.14	2.79	1.39
Rasogolla	2,000	200	0.40	0.15	0.13	2.10	1.70
Ghee (F–80%)	1,200	50	0.06	0.05	5.50	0.44	0.33
Total		300	6.05	4.52	163.20	49.83	38.27

Source : Aneja, Chandan, Banerjee (2002) : *op. cit.*, p. 226.

TABLE 3.18

Breakdown of Operating Surplus and Net Profit Per Year for a Dairy Plant in East India

Milk Product	Quantity Milk Processed (kg/day)	Product mftd. (No. of days)	Quantity Processed/ year ('000 tonnes)	Quantity Production/ year ('000 tonnes)	Sales per year (Rs. million)	Operating Surplus (Rs. million	Net Profit (Rs. million)
Dahi	5,000	100	0.50	0.49	15.68	3.20	2.20
Mishti Dahi	3,563	250	0.89	1.25	55.00	1.58	3.30
Lassi	5,000	60	0.30	0.37	9.28	3.84	3.24
Peda	5,150	100	0.52	0.15	15.40	3.14	2.14
Burfi/Kalakand	5,150	100	0.52	0.14	15.51	3.80	2.88
Gulabjamun	900	200	0.18	0.18	12.50	4.88	4.52
Kheer	1,000	50	0.05	0.40	2.40	0.91	0.82
Rabri	1,000	50	0.05	0.01	1.19	0.26	0.16
Paneer	10,000	100	1.00	0.22	19.80	4.14	2.14
Rasogolla	4,000	250	1.00	0.37	33.00	10.50	8.54
Sandesh	4,000	250	1.00	0.16	24.30	9.12	7.11
Ghee (F–80%)	1,200	50	0.06	0.05	5.50	0.44	0.33
Total		300	6.07	3.79	209.56	45.81	37.88

Source : Aneja, Chandan, Banerjee (2002) : *op. cit.*, p. 226.

BIS STANDARD

Listed below are Indian standards on dairy products, equipment, food additives, food hogging, testing method, etc. These standards can be purchased from the officers of the Bureaus of Indian Standards (BIS) Food and Agriculture Division Council (FADC), which have representatives of manufactures, consumers, government agencies, scientists and technologists.

Notes and References

1. *Indian Dairyman*, (2002), 54(2).
2. Vyas, B.M., *Indian Dairy Association*, West-Zone, p. 1.
3. Rao, Murlidhar, (1983), "Economic of Dairy Industry", *Kurukshetra*, April.
4. Nyholom, K., (1974), Socio-economic Aspects of Dairy Development, *Economic and Political Weekly*, 9 (52) A 127–A 136.
5. Khurody, D.N., (1974), *Dairying in India—A Review*, Asia Publishing House, Bombay.
6. Lampart, L.M., (1970), Modern Dairy Products, Asia Publishing House, Bombay.
7. V.M. Dandekar, (1969), *Economic and Political Weekly*, 4(39) 1559–1566.
8. Eckles, C. Henery, 2004, *Milk and Milk Products*, TMGH.
9. Datta, T.N., (2003), 'Economic Imperatives for Breed Improvement', pp. 31-38.
10. K.N. Nair, (1981), "Studies on India's Cattle Economy, *EPW*, 16(19), 21-23.
11. V.M. Rao, (1983), Towards White Revolution, *Kisan World*, 10(8), pp. 26-27.
12. Rageppa, S.R., (1977), "White Revolution in Karnataka, *Kurukshetra*, January).
13. Khuruddy, D.N., (1961), Economics of Small Milk Scheme; *Indian Dairyman*, Vol. 13, 16-18).
14. Rana, R.B.D. (1982), Dairies to Eradicate Rural Poverty, *Khadigramodyog*, 28(11), 504-10.
15. Rao, A.R., (1980), Social Cost of Employment in Dairying, *Eastern Economist*, 78(25), 1464-67.
16. Rao, C.K., (1966), *A Text Book of Dairy Science*, Calcutta.
17. Candler, W. and N. Kumar, (1998), The Dairy Revolution: The Impact of Dairy Development in India, World Bank.
18. Kurian, V., (1966), *Role of Milk in National Economy*, Agricultural Situation in India, 21(6) 455-64.
19. Elango, R. and S. Karthikegan, (1981), "Cost Analysis of Dairying", *Eastern Economist*; 77(19) 422–23.
20. Heady, E.O., (1964), Milk Production Function and Environment, *Journal of Farm Economies*, 46(i), 1-19.

21. Kenneth, N., (1954), *Dairy Farming : Theory and Practice*, London.
22. Kenney, R., (1957), *Dairy Husbandry*; New York.
23. Kurian, V. (1973) Rural Modernisation Through Dairy Development, *Kurukshetra*, October.
24. Delgado, E.M. *et. al.* (1999), Livestock to 2020 : The Next Food Revolution.
25. Punjrath, J.S. (1974), New Development in *Ghee* making, *Indian Dairyman*, 26(5), 275.
26. Aneji, *et. al.*, (1992), 'Traditional Milk Specialities—A Survey", *Dairy India*, 4th ed., Published by R.P. Gupta, New Delhi.
27. De, S. and Ray, S.C., (1953), "Studies on Indigenous Mehod of Khoa-making", *Indian Journal of Dairy Science*, No. 2(1), pp. 47-60.
28. Rajorhea, G.S. (1990), "Effects of Quality of Milk on Chemical Sensory and Rheological Properties of Khoa", *Indian Journal of Dairy Science*, Vol. 43, pp. 414-18.
29. Kalab, M. *et. al.*, (1988), "Development of Ultra Structure in Raw, Fried and Cooked Paneer made Buffalo, Cow and Mixed Milk", *Food Micro-structure*, Vol. 7, pp. 83-91.
30. Sachdeva, S. *et. al.*, (1985), "Recent Development in Paneer Technology", *Indian Dairyman*.
31. A.K. Adhikari, 1992.
32. De, S. and Ray, S.C. (1953), *op. cit.*, *Indian Journalof Dairy Science*, No. 17, pp. 113-15.
33. Patel, R.S., (1988), "Effects of Various Storage Temperatures on Sensory and Chemical Characteristics o Chakka", *Indian Journal of Dairy Science*.
34. Aneja, R.P., (1977), "Development of an Industrial Process for the Manufacture of Shrikhand", *Journal of Food Science Technology*, Vol. 14, pp. 159-63.
35. Banerjee, A.K. (1967), Dahi-making Equipment, *Indian Dairyman*, Vol. 19, No. 177-78.
36. Bhattacharya, D.C., (1967), New Varieties of Dahi, *Indian Dairyman*, Vol. 19, pp.35-38.
37. Siddhu, J.S. (1988), "Qualitative Aspects of Buffalo : Milk Constituents for Product Technology", Buffalo Production and Health, 2nd World Buffalo Conference, pp. 263-87.
38. Ghosh, J. and Rajorhea, G.S. (1990), *'Technology for Production of Misti Dahi—A Traditional Fermented Milk Product'*, Indian Journal of Dairy Science, Vol. 43, pp. 239-46.
39. Ramana, B.L.V. (1994), *'Indian Dairy Products'*, Asia Publishing House, New Delhi.
40. De, S. and Singh, B.P. (1970), Continuous Production of Khoa, Indian Dairyman, 22(12), p. 294.
41. Reddy, C.R. (1985), *'Process Modifications for Production of Khoa Based Sweets'*, Indian Journal of Dairy Science, Vol. 44(2), pp. 187-92.
42. Kumar, R. *et. al.* (1997), *'Shelf-life Extension of Peda using different Packaging Techniques'*, The Indian Journal of Dairy Science, Vol. 50(1), p. 40.
43. Ghosh, B.C. *et. al.* (1986), *'Complete Gulabjamun Mix'*, Indian Dairyman, Vol. 38(1986), p. 279.

44. Dharam Pal (2000) : *'Technological Advances in the Manufacture of Indian Milk Products—An Overview'*, Indian Dairyman, Vol. 52, No. 1, pp. 32-33.
45. Gayan, D. and Dharam Pal (1991) : *'Studies on the Manufacture of Storage of Rabri'*, Indian Journal of Dairy Science, No. 44, p. 80.
46. Swaminathan, M.S. (1990) : *'Indian Council of Agricultural Research'*, Hand Book of Animal Husbandary.
47. ICAR (1990) : Hand Book of Animal Husbandary, Publication Division, New Delhi, pp. 672-73.

4

Dairy Development in Five Year Plans

Livestock which constitutes an important portion of the wealth of the country is capable of making a large contribution to development of rural economy and thus, helps in raising the standard of rural masses. Agriculture contributes to the G.D.P. to an extent of 25% of which one-fourth is through livestock products. The development of livestock can be achieved through the following steps[1] :

(i) The increase in supply of milk, meat and eggs which are essential to balance the present poor custody diet.
(ii) The improvement in the drought animals to help the cultivator in his industry in a more efficient way.
(iii) The increase in the by-products such as wool, hair, hides and skin to raise their income.

Livestock production practices in India have been found to be not only labour-intensive but also labour-distributive and rural in Nature.

The National Commission on Agriculture has therefore identified the promotion of livestock production through the weaker sections of farmers in rural areas which could be relied upon as a major instrument of social change for supplementing the income and providing a large scope for employment for these sections of people in the rural areas.[2] Livestock is an integral part of agriculture dominated rural economy of Bihar, and it is firmly inter-woven with socio-economic matrix. It acts as agency for agriculture production, provides source for nutritive food materials, and raw materials for industries. It also helps in increasing crop production through making fluid cash available for purchase of critical agricultural inputs. Organic manure from cattle, poultry, sheep, goats or pigs has been essential for crop production, specially for vegetables and horticultural crops. Crop production alone cannot give adequate employment and income to bulk of the rural population. Further, it assumes special significance as land-man-ratio in the state is quite low. Fertile alluvial soils of north Bihar districts such as Begusarai, Muzaffarpur Darbhanga and other districts like Bhojpur, Buxar, Rohtas, Kaimur, Patna, Nalanda, etc. have good potential for dairy development in the state.

DAIRY DEVELOPMENT AND STEPS TAKEN DURING FIVE YEAR PLANS

Milk is the principal source of animal protein for majority of the population in India. In this country also proteins of high biological value are derived mainly from milch animal sources which are quite essential for infants, expectant and nursing mothers, old and infirm and manual workers.[3] The value of milk and milk products in the human diet is universally recognised and milk has occupied a prominent place in the dietary of people of Bihar from time immemorial.

In most of the states majority of population live in rural areas who are either cultivator or agricultural labourers. Very often, they are unemployed or under-employed due to vagaries of nature. The produce on the marginal land in various states often falls short of the dietary requirement of the farmer's household and seldom makes him indebted. Due to seasonal activities in agriculture in poor states the agricultural labourers are forced to remain unemployed and starve for want of alternative source of livelihood. Even after more than 50 years of Independence. The agricultural labourers are paid at the minimum rate which is not sufficient even to meet the minimum day-to-day requirements in Bihar.[4]

Dairy farming is a significant source of generating income and employment.[5] It provides supplementary income to the farmer households and utilisers the idle family members available in the farmers' families. Due to lack of alternative source of regular income, they are often indebted to meet the necessary farm expenditure. Instead income from milk is regular so that the farmers take the advantage of utilising the money from milk for the farm. Thus, dairy farming in Indian states is mostly practised as a subsidiary industry to agriculture. Dairy farming provides much more stable and continuous economic base than crop production and security against the vicissitudes of drought and famine in Bihar.

In Bihar also dairy farming development programmes are labour intensive, having favourable cost benefit ratio and are particularly suitable for weaker section of the rural community and have got redistribution effect in favour of them.

It offers more employment and income opportunities to small and marginal farmers and agricultural opportunities in Bihar state too because rising cattle and buffalo usually involves more intensive use of labour than many other enterprises. According to the Report of the National Commission of Agriculture, (1976, Vol. II, Animal Husbandry, Govt. of India). A very large proportion of female labour force finds scope for fuller utilisation in several operations connected with cattle and buffalo rearing.

In India, dairy farming apart from generating substantial direct employment, has also a large built in potential for generating indirect employment in several ancillary activities like manufacture of feed for livestock milk processing and manufacture of milk products, etc.

Besides, generating rural income and employment, milk protect human health by supplying rich proteins and nutrients and helps in increasing the standard of living of the rural masses. The milk animals also provide rich manure for farmers to raise farm productivity.

A well-organised dairy farming helps to increase milk production in rural areas by introducing of high yielding breed of cattle, and of better feeding and management practices enhancing the production resulting in better income levels to the rural population. Priority is sought to be given to the dairy development because white revolution of impact it has on the economy of the small and marginal farmers and agricultural labourers by way of generating larger employment opportunities and thus helping to increase their income[6] (Sixth Five Year Plan, 1980-85, Vol. 8, p. 63).

But the contribution of Bihar in Dairy farming in national level is quite poor. This has been the state of mainly due to low average yields of cows and buffaloes in the state of Bihar. The low milk production in Bihar is due to extremely low production potential of the animals and lack of adequate nutrition and health coverage. Only two-thirds of the fodder and one-fourth of the concentrates required for providing adequate nutrition to the present animal population are produced in Bihar. Consequently, most cows and she buffaloes, dry as well as wet and young stock are underfed in most of the districts of Bihar.[7]

The large gap between requirements availability of milk and milk products can be bridged by an aggressive cattle and buffalo development programmes, comprehensively organised and efficiently implemented.

In many states also there are many breeds of cattle and buffaloes, but the vast majority of these cattle and buffaloes are non-descript and poor productivity. The demand for milk and milk products is a function of dietary habits and food preferences, size of population and rural incomes. Consumers preferences for milk and milk products differ within and between regions. According to the Report of the National Commission on Agriculture, (Government of India) about 50% of the milk produced in this state is reported to have been consumed in fresh liquid form as modern facilities for moving fresh liquid milk have become available and as urban demand pressure has attracted milk production towards town and cities.

The gap between demand and supply of milk in cities continued to grow and acute shortages became patent in more and more areas. Widespread adulteration of milk with water and undesirable practice of urban milk production came into existence, and became a part of the general system of dairy farming.

In this country, because of shortages, price of milk and milk products continued to soar higher and higher taking these out of rich of the poorer sections of the community.

But as has been pointed out that the milk powder production capacities remain very much under utilised. The main criterion for evaluation of the economics of milk production is the cost and profitability of milk production. Under rural conditions with the producers own labour with level of milk production averaging 1,800 litres of milk per lactation and above, net profit to the producer will emerge. Dairying can't progress in this state unless milk production becomes an economic unit. Milk production can be an economic proposition only under following conditions :

(a) supply of nutrients required for milk production as cheaply as possible, and
(b) higher levels of milk production are obtained from the dairy cow so that the cost of milk production can be kept low.

In the words of D. Sunderesan (pp. 139-40) "In many dairying countries of the world due to concerted efforts to increase milk production over several decades, high yielding cattle have been produced. 'Holstein Friestein' cattle averaging 7000 litres of milk per lactation, 'Brown Swiss' and 'Rea Dane' cattle averaging 5000 to 6000 litres per lactation and Jersy cattle averaging 4000 litres of milk per lactation are available in many parts of the world. The advantages gained by these breeds could be exploited in just one generation by mating bulls from such breeds with Indian cattle. Several investigations spread over many institutions in India have produced high yielding crossbreed cattle through this method averaging 2500 to 4000 litres of milk per lactation. As crossbreed animals are bred among themselves there is a slight fall in milk production in further generations due to loss of heterosis."

The development of rapid implementation of artificial insemination programmes has been quite significant. The germ plasm from high yielding bulls is imported. Through frozen semen large scale utilization of bulls of higher quality, both Indian and imported breeds, is also resorted to through processing storage and utilisation of semen in liquid and frozen stage. Technology has been developed for diluting semen in proper media, and storing in liquid at about 5°C or freezing and storing in liquid nitrogen. The non-descript milk animals should be replaced gradually with cross bred animals so as to increase milk production and obviously the cost of production a per litre is to be minimised by producing higher quantity of milk.[8]

Feeds constitute a major item of cost of accounting for 60 to 80 percent of the cost of milk production, hence, obviously it is sought to be considered carefully in an attempt to reduce cost of milk production. In Bihar and other poor states also a rigorous effort should be made to supply feeds at a cheaper rate to the milk producers through the industry so that the cost of milk production is maintained.

The development of dairying depends on the price policy adopted by the dairy plants for the purchase of milk. The cost of milk production is to be taken into serious consideration while determining the price policy and any pricing policy not having such a base bound to erroneous. A just pricing structure should ensure that production of milk is encouraged and the farmers get lucrative returns for milk they supply. It is necessary that the purchase price should be such as to attract the inputs required for production such as labour and cultivable land, the use of which generally has a high opportunity cost.

The sufficient conditions for success include the competitiveness of the purchase price and of the timing and reliability of payment, the more competitive process can be reduced, the less of course is their impact on purchase prices.

The National Commission on Agriculture points out that only about 10% of the total milk handled by the organised dairying sector is cow milk, although cows accounts for 40% of milk produced in the country. This phenomenon can be attributed in part to the existing pricing systems that often favour the production of buffalo milk. While there is an advocacy for common pricing of cow and buffalo milk, a very large section of the dairy industry is paying for milk only on the basis of its fat content. A common pricing for cows and buffaloes may not however, prove helpful to provide incentive for production of cow milk as there is a prevailing tendency to dilute buffalo milk and pass it as cow milk.[9]

Pricing of dairy products should be fixed in a way, that would enable the industry to pay remunerative prices to the milk producers, cover the cost of collection, processing and distribution of milk and milk products and keep a fair margin of profit and yet make the prices of the commodities co-operatives.

The quality of dairy product should be well maintained as to face the competition of other unit.

STORAGE OF MILK

Milk drawn from a healthy cow is sterile but it contains bacteria that have entered the teat camel through teat opening. They are pushed out during milk process. The number of bacteria varies from animal to animal. The fore milk contains greater number of bacteria than strippings.

Milk gets easily contaminated with dirt, bacteria and odours. Milk furnishes an excellent medium for growth of bacteria, particularly when not properly cooled. They produce chemical changes rendering it unpalatable. Pathogenic bacteria can also very well multiply in milk.[10]

Thus, milk may serve as a medium for dissemination of infectious diseases. Hence, the quality and conditions of production of milk can be judged on the basis of microbial contents. Thus, great care in production and handling of milk is necessary to put it in the hands of consumers in a satisfactory conditions.

Milk should be removed to the milk house immediately after it is drawn because the container nation may also take place if it is left in the barn. The milk should then be strained into cows. If the cows are carefully milked, straining may not be necessary. It is impossible to strain bacteria out of milk.

However, it is desirable to filter the milk to remove hairs, particles of feed or bedding or dert, etc. that may have fallen into milk during the production. A simple service pad type strainer may be used for this purpose.

NECESSITY OF COOLING MILK BEFORE STORAGE

It is impossible to produce milk without some bacteria. Thus, efforts should be to prevent multiplication of the bacteria that have gained access. This can be achieved by cooling the raw milk.

According to 1940 census, Bihar Cattle Population was 15 million, and the state had an acute shortage of milk in those days. Cattle population rose to 18.6 million by 1951. During the period of First Five Year Plan, to achieve the objective of the cattle improvement programme, the work was taken under 'key village scheme'. Each centre was to consist of three or four villages having altogether about 500 cows over three years of age. In these areas breeding was to be strictly controlled and confined to three or four superior bulls specially marked out and maintained by the farmers for this purpose. The unapproved bull were to be removed other essential features of cattle development namely, maintenance of records of pedigree and milk production, rearing of calves, feeding and disease control, were to receive adequate attention at every centre. The technique of artificial insemination was to be utilised in these areas to accelerate progress and reduce the requirements of approved bulls. Every centre was to have one veterinary assistant surgeon, one milk recorder and three stockmen. About 150 artificial centres for 600 key villages were proposed to be established during the First Five Year Plan.[11]

DAIRY DEVELOPMENT DURING FIRST FIVE YEAR PLAN (1951-56)

In Bihar, per capita availability of milk was around 4.4 cences/day. As many as 26 schemes under animal husbandry and veterinary were designed and implemented with an expenditure of 1.28 crores during First Year Plan (1951-56). These schemes were classified under :

(i) cattle breeding,
(ii) goshala development,
(iii) poultry development,
(iv) sheep and wool development,

(v) veterinary services,
(vi) veterinary education and services,
(vii) dairy development, and
(viii) milk supply.

During First Five Year Plan in Bihar alongwith the expansion of Bihar Veterinary College at Patna, attempts were made to establish Pusa Cattle Farm. Besides, 13 key Village Blocks and 52 AI Centres were also set-up far the purpose. In Bihar, a rapid survey of milk production was conducted and co-operative Unions of Milk at Patna, Bhagalpur and Muzaffarpur were also organised to procure milk from rural areas and to distribute to the urban areas of state. Since then all these Co-operative Milk Union have been engaged in the procurement of milk for rural areas of the state and also in its distribution to urban people in all the regions.[12]

But during First Plan the position of dairy development was quite poor and unsatisfactory.

DAIRY DEVELOPMENT IN SECOND PLAN (1956-61)

Some constructive efforts were made during Second Five Year Plan for dairy development in State. In this Plan a provision of Rs. 4.07 crore was made for total outlay on dairying, milk supply and animal husbandry in Bihar. As per 1956 cattle census Bihar had a cattle population of 17.8 millions with a per capita availability was around 4.3 oz/day. In this Plan 11 key village blocks were added to existing infrastructure and a sum of Rs. 47.07 lakh was spent on dairy development in the state. Various Co-operative Milk Unions were given assistance in the form of equipment, vehicle, etc. Consequently Patna Co-operative Milk Union received maximum help, during Second Plan period. By 1961 Bihar State had 19.8 million of business with a per capita availability of 5 oz/day of milk. Milk production during this plan increased from 1.88 lakh tonnes in 1950-51 to 1.191 lakh tonnes in 1960-61 but the per capita/per day availability of milk remained constant as 133 gm. in 1950-51 and in 1960-61.[13]

As such, during this Plan the position dairy development in Bihar was like First Plan which was not upto satisfactory level.

DAIRY DEVELOPMENT IN THE THIRD PLAN (1961-66)

Though a number of steps were initiated in Bihar during the Third Five Year Plan (1961-66)[14] for dairy development but were quite inadequate and in impressive during Third Plan however following objectives were identified in the Third Plan :

(a) to improve goushalas for increased milk production and better bulls;
(b) to produce improved bullock for agriculture and to increases per capita availability of milk to 6 oz/day;
(c) to provide the condition of existing veterinary hospitals and dispensaries and to eradicate rinderpest from state; and
(d) to improve nutritional level of livestock by popularizing production of fodder grasses and by introducing methods of conservation of surplus grasses and developing pastures.

Intensive cattle development project were taken up at Patna and Barauni and there were as many as 43 key village blocks by the end of Third Five Year Plan in Bihar. Milk supply schemes at Gaya, Chilling Plant at Jehanabad and Maner and rural creamery at Barauni were commissioned. In this Plan a dairy plant was installed at Dumraown and of extension units were established in milk sheds. During Third Plan a sum of Rs. 3.98 crores were spent on dairy development in Bihar. In 1966 the Bihar State had 28.7 million of livestock with a per capita availability of 3.5 oz of milk/day. An amount of Rs. 1.72 crores were spent as animal husbandry and dairying during Annual Plans (1967-69) in this state. During 1967-69 Barauni Composed Milk Plant Dhanbad and Darbhanga milk supplies schemes were taken up in undivided Bihar.

DAIRY DEVELOPMENT DURING FOURTH FIVE YEAR PLAN (1969-74)

For the first time a dynamic and constructive effort was made during Fourth Five Year Plan (1969-74)[15] to give a new boost to dairy development and two separate sections were provided for the propose, i.e. (A) Animal Husbandry, and (B) Dairy Development.

During Fourth Five Year Plan following objectives were identified for the development of dairy farming in Bihar, i.e.

(a) to increase the production of milk and milk products;
(b) to increase the draught capacity of bullocks by improval breeding and better feeding;
(c) to provide adequate facilities for treatment of livestock and ensure protection by preventive measures;
(d) to ensure increased production of feeds and fodder;
(e) to train personnel in specialized fields; and
(f) to conduct research on livestock problems.

Consequent upon these programmes, in Bihar during Fourth Plan it was expected that milk production would be increased for 15.53 lakh tonnes to 16.10 lakh tonnes during 1969-74.

So far as dairy development and supply of milk are concerned, Fourth Plan had fixed following objectives for this in Bihar:

(a) completion of spill over dairy schemes,
(b) consolidation and expansion of the existing dairy schemes to meet increasing demand for milk in cities,
(c) continuation of schemes for training of personnel in dairying,
(d) intensifying rural extension work including organisation of milk collection and assembling centres for supply of milk to the dairy plants,
(e) formation of dairy development corporation to streamline administration and increase of efficiency of dairy projects,

(f) establishment of milk depots in industrial towns, and
(g) establishment of rural dairies for distribution of milk in small towns and supply of surplus milk to the dairy factories.

Thus, the Fourth Year Plan in Bihar emphasised consolidation and development of dairy projects undertaken during previous Plans.

For the first time in Bihar, the Bihar Dairy Development Corporation Limited (BSDDC) was registered in 1972 under Companies Act for rapid implementation and controlling dairy project in the state. The management of dairy plants were transferred to the Corporation. The Bihar State Dairy Corporation (BSDDC) was to develop Dairy Co-operatives both at the village as well as milk shed level on 'Anand pattern' in Gujarat and it was expected that the milk shed level co-operative would take over the entire infrastructure created in due source.

DAIRY DEVELOPMENT DURING FIFTH FIVE YEAR PLAN (1974-79)

The Fifth Plan land emphasis on increasing milk production and development of draught animals. Operation Flood I (OF-I) for implementation of programme was identified and 1000 litres per day capacity Feeder Balancing Dairy (FBD) and 100 MT per day Capacity Feed Plant (CFP) were also set-up. The Bihar State Dairy Development Corporation after recruitment and training of necessary staff positioned the procurement and input wing (P & I) from 1975. A Spear Head Team (SHT) was deputed from National Dairy Development Board from the same year for helping the BSDDC in organising and developing the co-operatives in the same year. Subsequently, during this Plan Bihar Government fell it necessary to request Dairy Board (NDDB—National Dairy Development Board) for taking over the infrastructure created on management basis. However, despite the progress in the initial years being encouraging, the programme, for obvious reasons could not achieve the goals for which it was established.[16]

The Fifth Plan laid emphasis on increasing milk production and development of drought animals and set following objectives for dairy development in Bihar :

(a) to build up adequate and strong infrastructure in the potential milk shed areas, urban areas and at the apex level of the management so that producer farmers could be benefited within the stipulated period, for increasing milk production and its marketing through organised dairies;

(b) to take action for the production of adequate quantities of basic technical inputs required for increasing milk production;

(c) to popularise practices of dairying among rural masses and to reach economic benefits to then by providing remunerative prices and various other basic technical inputs;

(d) to organised adequate number of healthy milk producers co-operative societies in the milk sheds and provide economic relief to the producers; to raise the installed capacities of the existing dairy plant and establish new dairy plants in the milk shed areas; and

(e) to set-up adequate number of milk marketing and distribution centres in the industrial towns.

By the end of Fifth Plan it was established that milk production would be 61 lakh litres. The total financial requirement for implementation of above programmes was estimated at Rs. 10 crores. Operation Flood Programme was launched in Bihar state in 1977 and the Bihar State Diary Development Corporation was made responsible for implementing the programme.

The project aimed at improving the milk marketing and increasing milk procurement and production in rural areas. Initially, it was taken up in Patna, Shahabad, Gaya and Saran. Efforts were made to strength and consolidate BSDDC during this period. By the end of Fifth Plan (1979-80), milk production increased 18.6 lakh tonnes from 17.27 lakh tonnes in 1973-74.

DAIRY DEVELOPMENT DURING SIXTH FIVE YEAR PLAN (1980-85)

It was Sixth Plan which changed the whole structure and functioning of dairy development in Bihar and set new examples. During the Sixth Plan (1980-85) Bihar was put under the mainstream of dairy farming named as *'White Revolution'*.

A total number of 992 veterinary hospitals and dispensaries were functioning in Bihar and the per capita availability of milk stood at 80 gm/per day in the state in the beginning of the Sixth Plan.[17]

The main objectives of the Sixth Plan for dairy development aimed as :

(i) increasing production of livestock products,
(ii) promotion of improved animal husbandry practices,
(iii) promotion of animal husbandry as a viable source of subsidiary income and occupation for the weaker section, and
(iv) spending Rs. 20.60 crore on dairy development.

The Sixth Plan proposed Rs. 20.60 crore for the purpose in Bihar. By 1984-85 the Plan envisaged a production target of Rs. 23.50 lakh tonnes of milk in Bihar. For this purpose a separate directorate of dairying was established as commissioner-*cum*-special secretary as its head during the Sixth Plan.

In terms of dairy development in Bihar 6 fluid milk plants at Barauni, Patna, Ranchi, Darbhanga, Bhagalpur and Bokaro were set-up and three rural dairy centres at Khagaria, Samastipur and Fatuah, 797 dairy co-operatives and 6 milk known were functioning in 1979-80.

The Sixth Plan advocated integrated approach to dairy development including production, collection, processing and marketing of milk and a two-tier dairy co-operative structure. Emphasis was also laid on supply of technical inputs and services in the milk sheds. A Dairy Science College was also proposed to cater to the training requirements of the state.

In order to replicate the Anand Pattern (of Gujarat) dairy co-operatives, Bihar State Co-operative Milk Producer's

Federation Ltd. (COMFED) was constituted in April 1983 in this Plan.

The task of COMFED was to make policy decisions and to co-ordinate activities of milk unions in procurement, processing and marketing of milk. The federation was expected to guide milk unions in Bihar in providing services like artificial insemination veterinary first aid, vaccination against disease, feeding and management in operation flood milk sheds. Operation Flood-II was launched in the 16 districts of Bihar during the Sixth Plan and an amount of Rs. 11.55 crores were spent on dairy development to achieve the above referred objectives.

DAIRY DEVELOPMENT DURING SEVENTH FIVE YEAR PLAN (1985-90)

At the beginning of the Seventh Five Year Plan (1985-90) in Bihar, according to 1982 livestock census, there were 22 million bovines, 9 ICDPs, 82 key village institutions, 22 key village centres, 42 key village Blocks, 10 large towns A1 centres, 6 key village extension centres, and 112 A1 centres functioning. The Seventh Plan fixed its objects in two categories—

(a) Objectives for Animal Husbandry.
(b) Objectives for Dairy Development Programmes.

The main objectives of the Seventh Plan for animal husbandry were fixed as :[18]

(i) to increase production of livestock products, such as milk eggs, meat and wool and also craft capacity for bullocks by intensification of controlled breeding programmes,
(ii) to consolidate and strengthen the existing infrastructural facilities for livestock development in state, and
(iii) to promote animal husbandry as a viable subsidiary source of income and occupation particularly by better feeding, proper management and animal health cover.

Major Objectives of Plan for Dairy Development Programme

The Seventh Plan (1985-90) aimed at :[19]

(i) to increase milk production by encouraging dairying as self-supporting and economically viable activities,
(ii) to extend operation Flood-II to 22 districts of Bihar,
(iii) to establish 4000 Anand Pattern Dairy Co-operatives under COMFED,
(iv) to increase milk production at the rate of 5.86 lakh liters per day,
(v) to establish and expand an incremental of 6.40 lakh litres per day processing capacity and 3.10 lakh of chilling capacity,
(vi) to construct/expand an incremental of 200 tonnes of balanced cattle feed/day,
(vii) to cover 6.5 lakh of milch animals under technical input programme and 1.60 milch animals under AI Programmes,
(viii) to expand and promote milk marketing in the major cities of Bihar at Patna, Ranchi, Bokaro, Jamshedpur, Gaya, Bhagalpur, Dhanbad, Muzaffarpur, Darbhanga, Biharsharif, Munger, Arrha, Chapra, Purnia, Katihar, Bermo, Danapur, Jamalpur, Daltonganj, etc.
(ix) to provide Rs. 14.45 crore as an outlay for the purpose.

The Seventh Plan spent Rs. 14.45 crore on dairy development of Bihar.[20]

DAIRY DEVELOPMENT DURING EIGHTH FIVE YEAR PLAN (1992-97)

The Eighth Plan played a key role in dairy development in Bihar. In 1991 per capita availability of milk was 95.2 gram/per day. A total number of 1258 veterinary hospitals and dispensaries including 44 mobile veterinary dispensaries were functioning in the state of Bihar. The Eighth Five Year Plan (1992-97) adopted a new cattle breeding policy and on the basis of available breeds and agro-climatic situation, Bihar State was divided into three zones :[21]

(a) North Gangetic Plains,
(b) South Gangetic Plains,
(c) Plateau of Chotanagpur and Santhal paraganas.

The main objectives for animal husbandry and dairy development in Bihar during the Eighth Plan announced as :

(a) to increase production of livestock products such as milk, eggs, meat and wool and drought capacity for bullocks by intensification of controlled breeding programme,
(b) to consolidate and strengthen existing infrastructural facilities for livestock development, and
(c) to promote animal husbandry as viable subsidiary source of income and occupation particularly better feeding, proper management and animal health cover.

At the end of Seventh Plan 2514 DCSs and 14 milk unions were registered. With the implementation of Operation Flood-II in Bihar, all dairy activities managed by BSDDC were handed over the COMFED and necessary funds were made available through budgetary allocation in this state.

Apart from the Operation Flood Programmes, Technology Mission on dairy development in state to create rural employment were also taken up during the Eighth Plan. An Integrated Dairy Development Programme was also taken up in the areas not covered under OFP. The Plan had a budget of Rs. 35.40 crores towards dairy development, and aimed at covering 40 towns with a total installed handling capacity of 14 lakh litres/day by organising 5050 DCSs in 20 milk sheds.

During this Plan the Feder Balancing Dairy with a capacity to handle 1.5 lakh liters per day (LLPD) had facilitated for manufacture of milk powder, butter, ghee, ice-cream, peda, paneer and plain/misti dahi. The production and marketing of table butter, under the brand name *Sudha* was introduced from 1st October, 1993 and the response was encouraging. The marketing of Sudha brand of ice-cream in Patna after test marketing in August-September 1994, was formally launched from April 1995. The initial response had been more than satisfactory.

DAIRY DEVELOPMENT IN THE NINTH FIVE YEAR PLAN (1997-2002)

At the beginning of the Ninth Five Year Plan (1997-02), there were 1258 veterinary hospitals and dispensaries including 44 mobile veterinary dispensaries in Bihar. The broad framework of the cattle and buffalo breeding policy being followed since the mid-sixties envisaged selective breeding of indigenous breeds in their breeding tracts and use of such improved breeds for upgrading of the non-descript stock. In the IXth Plan the contribution of the milk group alone Rs. 1,01,990 crore in India was higher than wheat Rs. 47,091 crore and sugarcane Rs. 27,647 crore. Milk production in India remained more or less stagnate from 1950-70. Thereafter it increased rapidly, reaching 84.6 million tonnes (mt.) in 2001-02. But the Ninth Plan target of milk production (94.49 mt.) was not achieved. The per capita availability of milk increased from 112 gm. per day in 1973-74 to about 226 gm. per day in 2001-02. However, it is still below the world average of 285 gm. per day. Investment in the dairy sector in the Ninth Plan decreased significantly compared to the Eighth Plan. Out of 168 Milk Unions, 58 Milk Unions (34.5%) were running in loss in 2000 in India.[22]

So far as the Government policy in the dairy sector has been to give preference to the establishment of milk processing plants linking rural milk producers to urban consumers through a network of co-operatives. Restrictions establishing new milk processing capacity under milk and milk products order (MMPO) has now been removed. No policy measures have been undertaken so far to give a fillip to the unorganised sector involved in the production of Indian dairy products like *ghee, paneer, khoa,* etc. which have tremendous potential in the export market in Asian and African countries.

The main objectives for Animal Husbandry in Bihar in the Ninth Plan were target under :

(a) to increase livestock production such as eggs, meat, wool and also drought capacity for bullocks by intensification of controlled breeding programme,

(b) to consolidate and strengthen the existing infrastructural facilities for livestock development in Bihar, and

(c) To promote husbandry as a viable subsidiary source of income and occupation particularly by improved breeding, feeding, proper management and animal health cover.

The Ninth Plan aimed at among other strengthening government of cattle farms, continuation of frozen semen banks, etc. The plan had a target of 40.25 lakh tonnes of milk production by 2002 with a total budgetary allocation of Rs. 38.26 crores.

Major Objective of Dairy Development Programme in 9th Plan

The major objectives of dairy development programme during the Ninth Plan were to increase milk production by encouraging dairying as self-supporting and economically viable activity for small and marginal farmers and rural landless labourers.

The target of the Ninth Plan was to cover 40 towns inclusive of 23 towns of Operation Flood Programme with total installed handling capacity of 7.33 lakh litres/day. By the end of 1999-2000, 3700 milk producer's co-operative societies were registered.

During this Plan Feeder Balancing Dairy has been promoted with a capacity of a 2 lakh litres per day which had the facilities of milk powder, butter, ghee, ice-cream, peda, paneer and plain misti/dahi. The production and marketing of milk products under the brand name of Sudha have been promoted. Similarly with certain modification in the dairy policy decisions and because of concerted efforts the quantum of milk marketed is steadily growing. The Dairy Plant Management Programme (DPM) introduced in the year 1992 followed by Quality Assurance Programme (QAP) in 1993 with the help of NDDB were quite successful during the plan period.

DAIRY DEVELOPMENT DURING TENTH FIVE YEAR PLAN

Animal husbandry and dairying have received high priority in the efforts for generating wealth and employment, increasing the availability of animal protein in the food basket and for generating exportable surplus.[23]

The overall focus during Tenth Plan is on four broad pillars, viz.:

(i) removing policy distortion that is hindering the natural growth of livestock production,

(ii) building participatory institutions of collective action for small scale farmer that allow them to get vertically integrated with livestock processors and inputs suppliers,

(iii) creating an environment in which farmers will increase investment in ways that will improve productivity in the livestock sector, and

(iv) promoting effective regulatory institutions to deal with the threat of environmental and health crises stemming for livestock.

The Tenth Plan target for milk production in India is set at 108.4 mt. envisaging an annual growth rate of 6 percent.

In respect of dairy development, Tenth Plan (2002-07), had the following objectives :

(a) to increase livestock production such as milk, eggs, meat and wool and also drought capacity of bullocks by intensification of controlled breeding programme,

(b) to consolidate and strengthen existing infrastructural facilities for livestock development in the state, and

(c) to promote animal husbandry as a viable subsidiary source of income and occupation of improved breeding, feeding, proper, management and animal health cover.

For this purpose an amount of Rs. 39.70 crore has been proposed. A new cattle breeding policy has been adopted with pure Jersey or Friesian with a view to maintain 50% exotic inheritance in both North and South Gangetic plains. All the centrally sponsored diary schemes, survey on milk and establishment of low input technology are being continued in the present Tenth Plan.

The main objective of dairy development programme during the Tenth Plan in Bihar is to increase milk production by encouraging dairying a self-supporting and economically viable activity. The total plan allocation for dairy development in Bihar has been estimated it Rs. 8.50 crores.

Thus, during various Five Year Plans dairy sector has achieved glorious position. Patna Dairy and *Sudha* have done tremendous job and have performed quite well and promoted *'White Revolution'* in Bihar. COMFED in Bihar is the brain child of Mr. V. Kurien the *God father* of *White Revolution* in India and who made globally renowned effort apply called, Operation Flood. Patna joined the Revolution then and earlier in its own peculiar brand. Patna was the only city in India, perhaps where the practice of bringing the cow all the way the consumer's gate to be milked under his watchful eyes. The development of *'Sudha'* is a new strategy to satisfy the consumers of Bihar through making available the improved quality of milk products on reasonable price in sufficient supply and to meet the requirements of milk and milk product in a quite systematic, sustainable, sufficient and scientific manner. During Tenth Plan *Sudha* has been declared as the market leader in the consolidated scheme of Brand development in Bihar.

Sudha has been recognised as the glory of Bihar with the total turn over of Rs. 450 crore per annum Sudha is one of the successful stories of milk trending. It has given a right track/ direction to the 'Relationship Marketing approach'. COMFED in Bihar has established a long-term business relationship instead of short-term business relationship with the valuable and honourable consumers, milk producers, distributors and dealers in the State during recent years. The main goal of Federation and this institution is to provide long-term-based profits and gains to the consumers and to establish a strong chain between Brand and the consumers. This type of stronger relationship can be established only through availability of high level improved quality of milk and milk product on reasonable price and better and harmonious services.[24]

In this sector, four aspects of dairy marketing a centrified effort has been made in Bihar by COMFED after conducting intensive studies of consumer's behaviour under various marketing sections.[25] This institution has been making consistent effort to implement total quality management by improving quality process in every stage of dairy and dairy milk product and dairy operation. Efforts are being made to satisfy consumers' are recent and new requirements in changed situation by developing new products and improving and expanding milk products.[26] Some new products have also been developed in view of making remarkable presence. Sudha is ready to provide milk to all with high and improved quality of milk. Sudha dairy has been providing Lassi, Rasgulla, Peda, Burfi, Misti, Gulabjamun at reasonable and competitive price in open market to the consumers. It has left very good impression in the market.

Obviously, India is the largest milk producing country of the world with a share of about 14% in world milk production. Milk has achieved a unique status in terms of the output and contribution to the national economy. With output value exceeding Rs. 1,00,000 crore and has made rapid strides both in terms of number of milk products and the quantity of milk produced. Over the last two decades, while both population and foodgrain production grew at around 2% milk production grew at more than double the rate of growth of the population, increasing per capita availability of milk from 112 gms/day in 1970-71 to 231 gms/day in 2003-04.

A growth rate of 4.5% has been achieved by the dairy sector during the past decade as compared to the 2% growth recorded by the agricultural sector as a whole. In fact, an annual production of 1200 million tonnes of milk can be achieved by the year 2010, only of the growth rate is kept at a rate not less than 5.5%. The value of milk output and its products Rs. 70,000 crore and that as dairy industry as a whole Rs. 1,05,000 crore.

As per World Bank experts for an initial investment of Rs. 200 crore in Operation Flood II the net return/year to the rural economy had been 24,000 crore Rupees. No other major development programme all over the world has matched this input-output ratio.[27] (*Kurukshetra*, Dec. 2005, pp. 30-35)

TABLE 4.1

Consuption (Per Capita and Total) by 4 Milk Groups in 2001

Levels of Income	*Population (million)*	*Per Capita Milk Consumption*		*Total Milk Consumption*
		kg/year	*gm/day*	*(MT)*
Rural India				
I	111.30 (15%)	6.6	18.2	0.75 (01.9%)
II	178.08 (24%)	20.9	57.4	3.73 (09.6%)
III	222.60 (30%)	43.4	119.0	9.67 (24.8%)
IV	230.02 (31%)	108.2	296.2	24.86 (63.7%)
Total	742.00 (100%)	52.6	144.0	39.00 (100%)
Urban Area				
I	37.05 (13%)	45.1	123.6	1.67 (03.7%)
II	71.25 (25%)	86.4	236.8	6.16 (13.5%)
III	91.20 (32%)	161.1	441.4	14.69 (32.2%)
IV	85.50 (30%)	269.9	739.6	23.08 (50.6%)
Total	285.00 (100%)	160.0	438.3	45.60 (100%)
All India				
I	148.35 (14.4%)	16.1	44.3	2.40 (2.8%)
II	249.33 (24.2%)	39.6	108.6	9.88 (11.7%)
III	313.80 (30.5%)	77.6	212.7	24.36 (28.8%)
IV	315.52 (30.7%)	151.9	416.3	47.94 (56.7%)
Total	1028.00 (100%)	82.4	225.7	84.60 (100%)

Income Level : I—Very Poor, II—Moderately Poor, III—Non-poor lower, IV—Non-poor higher.

Source : Based on data published in Food Demand and Supply projecions for India by Prof. Praduman Kumar, Agricultural Economics Policy, Paper No. 90-01, Indian Agriucltural Research Institute (IAR), New Delhi, 1998.

TABLE 4.2

Population Projection—Rural and Urban—by Socio-economic Groups, 2001 to 2016

(*Million*)

Region	*Year*	*Very Poor*	*Poor*	*Non-poor Low*	*Non-poor High*	*All Groups*
(1)	*(2)*	*(3)*	*(4)*	*(5)*	*(6)*	*(7)*
Rural India	2001	44.5	111.6	252.6	314.1	722.8
	2006	31.9	103.0	264.6	365.3	764.8
	2011	20.9	91.1	270.5	421.7	804.2
	2016	14.7	82.0	273.7	470.2	840.6
Eastern	2001	22.6	46.1	77.0	51.8	197.5
	2006	17.0	44.7	84.7	63.4	209.8
	2011	11.62	41.6	91.4	77.8	222.3
	2016	8.4	38.7	96.4	91.3	234.8
Western	2001	9.9	26.1	63.8	85.2	184.9
	2006	6.7	22.9	65.7	98.9	194.3
	2011	4.1	18.9	65.7	127.7	211.6
	2016	2.8	16.1	65.1	127.7	211.6
Northern	2001	8.6	25.4	56.6	56.9	167.6
	2006	6.5	24.7	60.8	89.5	181.4
	2011	4.5	23.1	64.4	104.3	196.3
	2016	3.3	21.9	67.7	118.6	211.5
Southern	2001	3.1	12.6	48.2	86.3	150.2
	2006	1.6	9.5	46.5	96.4	154.0
	2011	0.65	6.6	43.1	106.4	156.7
	2016	0.31	4.8	39.7	113.6	158.3
Hills	2001	0.12	0.54	3.6	9.4	13.6
	2006	0.07	0.37	3.16	11.16	14.8
	2011	0.03	0.21	2.44	12.4	15.0
	2016	0.02	0.12	1.82	12.4	14.4
Northern-eastern	2001	0.17	0.87	3.39	4.48	8.9
	2006	0.11	0.81	3.81	5.89	10.6
	2011	0.05	0.59	3.50	6.56	10.7
	2016	0.03	0.42	2.98	6.56	9.9

(*Contd.*)

TABLE 4.2 (Contd.)

(1)	(2)	(3)	(4)	(5)	(6)	(7)
Urban India	2001	20.9	36.5	72.6	163.8	293.8
	2006	16.9	34.6	76.7	206.7	334.9
	2011	12.7	31.1	78.2	258.2	380.2
	2016	10.1	28.5	79.9	310.5	429.0
Eastern	2001	3.4	6.6	13.2	24.9	48.4
	2006	2.8	6.4	14.2	30.8	54.2
	2011	2.2	6.0	14.8	38.0	61.1
	2016					68.6
Western	2001	9.4	13.8	26.4	45.4	94.9
	2006	8.1	13.5	28.8	57.2	107.6
	2011	6.5	12.6	30.5	71.7	121.3
	2016					136.3
Northern	2001	4.1	7.9	16.7	38.9	67.5
	2006	3.4	7.7	18.3	50.3	79.7
	2011	2.5	7.1	19.3	64.7	93.5
	2016					136.3
Southern	2001	4.0	8.0	15.5	48.5	76.1
	2006	2.7	6.9	14.8	60.3	84.7
	2011	1.6	5.4	13.1	73.5	93.6
	2016					108.9
Hills	2001	0.003	0.064	0.41	2.91	3.38
	2006	0.001	0.036	0.29	3.65	3.98
	2011	0.000	0.026	0.17	4.33	4.52
	2016					4.96
North-Eastern	2001	0.012	0.026	0.42	3.3	3.72
	2006	0.009	0.016	0.37	4.47	4.87
	2011	0.005	0.007	0.29	5.8	6.13
	2016					7.39
All India	2001	65.4	148.1	325.2	477.9	1016.6
	2006	45.8	137.6	341.3	572.0	1099.7
	2011	33.6	122.2	348.7	679.9	1184.4
	2016	24.8	110.5	353.6	780.7	1269.6

Source : Division of Agricultural Economics, Indian Agriculture Institute (IARI), New Delhi.

TABLE 4.3

Per Capita Quantity and Value of Monthly Comsumption, Rural and Urban of Liquid Milk and Milk Products, State-wise, 1999-2000

	Rural			Urban		
	Liquid Milk		Milk and Milk Products	Liquid Milk		Milk and Milk Products
	Qty. (Litres)	Value (Rs.)	Value (Rs.)	Qty. (Litres)	Value (Rs.)	Value (Rs.)
(1)	(2)	(3)	(4)	(5)	(6)	(7)
Andhra	2.87	26.19	27.40	4.40	49.08	53.15
Arunachal	0.50	6.10	16.30	1.89	21.69	39.15
Assam	1.11	13.45	15.05	2.14	30.76	43.07
Bihar[1]	2.41	23.62	25.77	3.40	42.61	47.70
Goa	3.16	44.47	47.65	4.22	62.44	73.48
Gujarat	5.42	66.95	78.60	6.58	88.23	110.88
Haryana	13.88	141.51	164.46	9.03	120.04	147.09
Himachal	7.87	84.00	95.29	10.08	119.36	151.21
J & K	9.53	86.23	93.91	8.02	85.18	111.78
Karnataka	3.45	41.57	33.75	5.07	53.78	61.70
Kerala	2.97	35.91	37.88	3.49	44.34	49.27
M.P.[2]	2.71	27.39	31.97	4.33	51.76	63.08
Maharashtra	2.66	28.69	29.93	4.79	66.02	72.61
Manipur	0.32	3.33	5.46	0.45	5.43	13.98
Meghalaya	0.92	8.81	10.00	2.92	30.88	39.42
Mizoram	0.44	5.35	18.79	1.54	22.32	45.06
Nagaland	0.86	11.13	36.58	1.78	23.48	55.97
Orissa	0.64	6.19	7.81	1.97	21.48	29.23
Punjab	11.67	118.60	127.90	9.73	115.68	129.93
Rajasthan	9.62	95.91	109.36	7.72	87.99	125.16
Sikkim	4.27	40.30	45.62	5.97	52.83	61.68

(Contd.)

TABLE 4.3 (*Contd.*)

(1)	*(2)*	*(3)*	*(4)*	*(5)*	*(6)*	*(7)*
Tamil Nadu	2.39	24.21	25.22	4.77	53.59	57.95
Tripura	1.32	16.75	19.92	3.09	31.86	46.41
U.P.[3]	4.52	43.48	46.66	5.27	62.23	72.05
West Bengal	1.31	13.13	14.65	2.63	30.68	40.31
Union Territories						
A & N Island	1.10	13.84	38.60	1.29	22.00	62.57
Chandigarh	9.92	131.52	144.96	10.53	141.24	172.96
D & N Haveli	1.43	19.19	21.53	5.49	83.94	101.48
Daman & Diu	3.13	45.16	55.15	5.56	77.34	98.69
Delhi	6.34	91.32	133.02	8.73	122.43	154.18
Lakshdweep	0.36	5.63	29.30	0.55	8.51	28.35
Pondicherry	2.65	27.52	29.62	4.64	48.64	52.47

[1]Includes Jharkhand; [2]Includes Chhattisgarh; [3]Includes Uttaranchal

Source : Consumption of some important commodities in India, 1999-2000, National Sample Survey 55th Round (July 1999-June 2000), Ministry of Statistics and Programme Implementation, Government of India.

TABLE 4.4

Per Capita Quantity and Value of Monthly Comsumption of Liquid Milk and Milk Products by Monthly Per Capita Expenditure (MPCE) Class, 1999-2000

MPCE Class	*Liquid Milk*		*Total Value of Milk and Milk Products (Rs.)*
	Qty. (Litres)	*Value (Rs.)*	
(1)	*(2)*	*(3)*	*(4)*
Urban India			
000-300	1.27	13.02	13.64
300-350	1.68	18.89	20.68
350-425	2.21	25.17	27.37

(*Contd.*)

TABLE 4.4 (Contd.)

(1)	(2)	(3)	(4)
425-500	2.86	32.79	36.25
500-575	3.42	39.90	45.24
575-665	4.33	50.53	57.32
665-775	5.15	59.15	68.50
775-915	5.78	70.36	83.63
915-1120	6.79	85.29	101.94
1120-1500	7.91	100.93	121.54
1500-1925	10.04	129.01	158.65
Above 1925	12.04	162.35	204.75
All Classes	5.10	62.66	74.17
Rural India			
000-225	0.40	3.84	4.06
225-255	0.68	6.58	6.94
255-300	1.07	10.06	11.07
300-340	1.61	15.41	16.30
340-380	2.12	21.13	22.64
380-420	2.63	25.97	27.93
420-470	3.30	32.96	35.99
470-525	4.39	40.31	44.45
525-615	5.15	52.17	57.74
615-775	6.83	70.43	78.28
775-950	8.77	91.76	103.68
Above 950	11.99	129.76	150.01
All classes	3.79	38.37	42.56

Source : Consumption of some important commodities in India, 1999-2000, National Sample Survey 55th Round (July 1999-June 2000), Ministry of Statistics & Programme Implementation, Government of India.

TABLE 4.5

Per Capita Quantity and Value of Monthly Consumption of Six Milk Products by MPCE* Class, 1999-2000

MPCE Class	Baby Food		Condensed Milk/Powder		Curd		Ghee		Butter		Ice Cream
	Qty. (kg.)	Value (Rs.)	Qty. (kg.)	Value (Rs.)	Qty. (kg.)	Value (Rs.)	Qty. (kg.)	Value (Rs.)	Qty. (kg.)	Value (Rs.)	Value (Rs.)
(1)	(2)	(3)	(4)	(5)	(6)	(7)	(8)	(9)	(10)	(11)	(12)
Urban Area											
000-300	—	—	—	—	—	—	—	—	—	—	—
300-350	—	—	—	—	—	—	0.01	0.87	—	—	—
350-425		—				—	0.01	1.11	—	—	—
425-500	—	—	0.02	0.55	—	—	0.03	2.09	—	—	—
500-575		—	0.03	0.77	0.04	0.55	0.03	3.51	—	—	—
575-665	—	—	0.04	0.57	0.04	0.62	0.04	4.93	—	—	—
665-775	0.03	0.76	0.02	0.85	0.06	0.80	0.05	6.46	—	—	—
775-915	0.01	0.72	0.02	1.05	0.07	0.97	0.07	9.74	—	—	—
915-1120	0.02	0.88	0.06	1.15	0.08	1.25	0.09	12.10	0.01	0.51	—
1120-1500	0.02	1.19	0.03	1.14	0.08	1.21	0.11	14.93	0.07	1.05	—
1500-1925	0.03	1.34	0.02	1.10	0.12	1.88	0.18	21.54	0.07	1.75	—

(Contd.)

Table 4.5 (Contd.)

(1)	(2)	(3)	(4)	(5)	(6)	(7)	(8)	(9)	(10)	(11)	(12)
Above 1925	0.02	1.20	0.05	1.48	0.12	2.04	0.19	26.98	0.11	5.03	4.14
All classes	0.01	0.60	0.03	0.81	0.06	0.83	0.07	7.98	0.02	0.58	—
Rural Area											
000-225	—	—	—	—	—	—	—	—	—	—	—
225-255	—	—	—	—	—	—	—	—	—	—	—
255-300	—	—	—	—	—	—	—	—	—	—	—
300-340	—	—	—	—	—	—	—	—	—	—	—
340-380	—	—	—	—	—	—	0.01	0.82	—	—	—
380-420	—	—	—	—	—	—	0.01	1.30	—	—	—
420-470	—	—	—	—	—	—	0.02	2.10	—	—	—
470-525	—	—	—	—	—	—	0.04	2.92	—	—	—
525-615	—	—	—	—	0.04	0.59	0.03	3.94	—	—	—
615-775	—	—	0.07	0.53	0.05	0.68	0.04	6.01	—	—	—
775-950	—	—	0.04	0.77	0.07	0.91	0.08	9.32	—	—	—
Above 950	0.01	0.57	0.06	1.54	0.09	1.50	0.18	15.41	0.02	0.65	—
All clases	—	—	—	—	—	—	0.03	3.03	—	—	—

* MPCE : Monthly per capita expenditure.

Source : Consumption of some important commodities in India, 1999-2000, National Sample Survey 55th Round (July 1999-June 2000), Ministry of Statistics & Programme Implementation, Government of India.

TABLE 4.6
Share of Expenditure on three Major Groups of Foods in Total Food Expenditure, Rural and Urban, 1972-73 to 1999-2000

Year	*Expenditure On*					
	Foodgrains		*Fruits/Vegetable*		*Milk, Meat, Egg and Fish*	
	Rural	*Urban*	*Rural*	*Urban*	*Rural*	*Urban*
1972-73 (27th round)	63.10	42.02	10.15	9.92	13.44	19.53
1977-78 32nd round)	58.01	40.83	9.95	10.67	16.17	21.67
1983 (38th round)	55.34	38.75	11.43	12.01	16.01	21.66
1987-88 (43rd round)	47.81	33.16	12.66	13.83	18.59	23.23
1993-94 (50th round)	44.78	31.63	14.40	14.99	20.25	24.13
1999-2000 (55th round)	44.11	31.81	14.48	15.59	20.37	24.53

Source : National Sample Survey Reports, Ministry of Statistics and Programme Implementation, Government of India.

TABLE 4.7
Trends in Annual Milk Production and Per Capita Availability, State-wise, 1999 to 2002

State/Union Territory	*Milk Porduction ('000 tonnes)*			*Per Capita Availability 2001-02*	
	*1999-2000**	*2000-2001**	*2001-2002***	*kg/ year*	*grams/ day*
(1)	*(2)*	*(3)*	*(4)*	*(5)*	*(6)*
Andhra Pradesh	5,122	5,521	5,145	67.9	186
Arunachal Pradesh	45	45.5	55	5.0	14
Assam	733	738	894	33.6	92

(Contd.)

TABLE 4.7 (Contd.)

(1)	(2)	(3)	(4)	(5)	(6)
Bihar[1].	3,740	3,878	4,068	37.1	10
Goa	43	44	47	34.97	96
Gujarat	5,255	5,317	5,573	110.1	302
Haryana	4,679	4,849	4,976	236.0	647
Himachal Pradesh	741	760	810	133.3	365
Jammu & Kashmir	1,286	1,037	1,088	108.4	296
Karnataka	4,473	4,598	5,357	101.6	278
Kerala	2,673	2,771	2,907	91.3	250
Madhya Pradesh[2]	5,600	5,806	6,091	75.0	206
Maharashtra	5,706	5,850	6,204	62.3	171
Manipur	67	69	73	30.6	84
Meghalaya	62	64	71	30.8	84
Mizoram	18	14	11	12.3	34
Nagaland	49.5	60	54	27.2	74
Orissa	847	875	865	23.6	64
Punjab	7,700	7,984	8,375	344.8	945
Rajasthan	7,260	7,455	6,330	112.1	307
Sikkim	35	35.5	46	85.1	233
Tripura	49	51	53	16.6	46
Uttar Pradesh[3]	14,153	14,840	16,506	94.6	259
West Bengal	3,465	3,470	4,079	50.8	139
Union Territories					
Andman & Nicobar Island	23	24	25	70.1	192
Chandigarh	42	44	46	51.1	140
Dadar & Nagar Haveli	10	10	11	69.6	191
Daman & Diu	1	1	1	22.7	62
Delhi	292	292	321	23.3	64
Lakshdweep	1	1	1	16.5	45
Pondicherry	36	37	38	39.0	107
All India	78,779	81,430	84,570	82.3	226

* Provisional ** Anticipated

[1]Includes Jharkhand; [2]Includes Chhattisgarh; [3]Includes Uttaranchal

Source : Department of Animal Husbandry and Dairying, Ministry of Agriculture, Government of India.

TABLE 4.8
Pockets of Major Milk Production in India, Region-wise and State-wise, 2001

('000 litres per day)

Northern Region
(101,000)*

Delhi (900)*

Gazipur, Jharoda, Kakrola, Madanpur, Nangloi.

Haryana (13,600)*

Hisar (1,783), Rohtak (1,167), Jind (1,078), Karnal (716), Kurukshetra (667), Kaithal (627), Ambala (695).

Punjab (23,000)

Amritsar (2,517), Sangrur (2,071), Ludhiana (1,973), Patiala (1,843), Jalandhar (1,811), Ferozepur (1,652), Faridkot (1,543), Gurdaspur (1,478), Hoshiarpur (1,245), Bhatinda (918).

Rajasthan (17,300)

Jaipur + Dausa (1,708), Sri Ganganagar + Hanumangarh (1,497), Alwar (876), Sawai-Madhopur (873), Udaipur & Rajasamand (767), Sikar (733), Bikaner (652), Nagaur (600), Ajmer (571), Jhunjhunu (537), Churu (512), Jodhpur (410).

Uttar Pradesh (45,200)

Meeru (1,865), Aligarh (1,440), Moradabad (1,269), Allahabad (1,011), Agra (914), Varanasi (847), Mathura (795), Bereilly (654), Fatehpur (609), Gorakhpur (485), Lucknow (368).

Western and Central Region
(50,000)

Gujarat (15,300)

Mehsana (2,015), Kheda (1,584), Banaskantha (1,238), Ahmedabad & Gandhinagar (1,119), Sabarkantha (1,016), Junagadh (1,001), Bhavnagar (994), Rajkot (813), Panchmahai (802), Vadodara (775), Kutch (516).

Madhya Pradesh (16,700)

Dewas-Indore-Ratlam (1,252), Kolhapur (1,094), Satara (980), Jalgaon (853), Solapur (741), Greater Mumbai (706), Nashik (647), Nanded (562), Dhule (493), Parbhani (471), Beed (468).

TABLE 4.8 (*Contd.*)

Southern Region
(51,000)

Andhra Pradesh (14,100)

Guntur (1,105), West Godavari (1,048), Chittor (1,042), Krishna (939), East Godavari (869), Prakasam (844).

Karnataka (14,700)

Belgaum (1,292), Kolar (1,133), Mysore (1,091), Dharwad (1,071), Bangalore (1,014), Mandya (796), Tumkur (694), Gulbarga (611).

Kerala (8,000)

Thriruvananthapuram (1,058), Ernakulam (847), Kollam (820), Trissur (683), Kottyam (662), Palghat (614).

Tamil Nadu (12,700)

North Arcot + Triruvannamalai Sambuvarayar (1,209), Madurai + Dindigul Anna (1,042), South Arcot + Villupuram Ramaswamy Padaiyatchiar (1,028), Chengai MGR (1,014), Trichy (965), Salem (953).

Eastern Region
(30,000)

Bihar (12,000)

Begusarai, Vaishali, Patna, Saran, Bhojpur, Samastipur, Muzaffarpur, Sitamarhi, Khagaria, Rohtas, Saharsa, East Champaran, Gaya, Monghyr (Munger).

West Bengal (12,000)

Kolkata (3,279), Burdwan (1,289), Midnapore (1,231), Nadia-Kalyani (1,076), Murshidabad (664), Bankura (462), Darjeeling (384), Jalpaiguri (376).

Orissa (2,500)

Assam & North-Eastern States (3,500)

*Figures in brackets represent daily milk production in thousand litres.

Source : State Department of Dairy Development, Animal Husbandry and Veterinary Services, State Co-operative Dairy Federation, etc.

TABLE 4.9
Performance of the Anand-model Dairy Cooperatives and Milk Marketing in Four Metro Cities, 2000-01*

Region/State	*Dairy Cooperative societies*	*Farmer members ('000)*	*Average milk procure-ment ('000 kg/ day*	*Liquid milk marketing ('000 ipd)*
(1)	*(2)*	*(3)*	*(4)*	*(5)*
Northern Region				
Haryana	3,318	185	276	108
Himachal Pradesh	288	20	24	20
Jammu & Kashmir	—	—	—	—
Punjab	6,823	370	912	420
Rajasthan	5,900	436	887	540
Uttar Pradesh	15,648	649	791	436
Delhi	—	—	—	1,524
Total	31,977	1,660	2,890	3,048
Western Region				
Goa	166	18	32	83
Gujarat	10,679	2,147	4,567	1,905
Madhya Pradesh	4,877	242	319	244
Maharashtra	16,724	1,398	2,979	1,178
Mumbai	—	—	—	1,390
Total	32,446	3,805	7,897	4,800

(Contd.)

TABLE 4.9 (Contd.)

(1)	(2)	(3)	(4)	(5)
Southern Region				
Andhra Pradesh	4,912	702	879	733
Karnataka	8,516	1,528	1,887	1,510
Kerala	2,781	637	646	640
Pondicherry	92	27	45	43
Tamil Nadu	8,369	1,957	1,618	559
Chennai	—	—	—	725
Total	24,670	4,851	5,075	4,201
Eastern Region				
Assam	124	1	3	7
Bihar	3,525	184	330	324
Nagaland	74	3	3	4
Orissa	1,412	111	94	98
Sikkim	174	5	7	7
Tripura	84	4	1	7
West Bengal	1,719	114	204	27
Kolkata	—	—	—	840
Total	7,113	422	642	1,314
Grand Total	96,206	10,738	16,504	13,363

* Figures are provisional and include conventional societies and Taluka unions formed earlier.

— Data not available.

Source : National Dairy Development Board Annual Report, 2000-01.

TABLE 4.10

State-wise Population of Cows (Crossbreed and Desi) and Buffaloes, their Average Daily Yield and Annual Milk Production, 1998-99

State	In-milk Animals ('000 numbers)				Average yield Animal (kg./day)				Annual Milk Prod. ('000 tonnes)			
	Cow			Buffalo	Cow			Buffalo	Cow			Buffalo
	Cross Breed	Desi	Total		Cross Breed	Desi	Total		Cross Breed	Desi	Total	
(1)	(2)	(3)	(4)	(5)	(6)	(7)	(8)	(9)	(10)	(11)	(12)	(13)
Andhra Pradesh	165 (220)	1,305 (2,251)	1,47() (2,471)	4,612 (4,462)	6.571	1.80	2.36	2.939	407	859	1,266	3,577
Arunachal Pradesh	11.59 (15.27)	26.64 (40.85)	38.23 (56.12)	—	7.67	1.30	3.22	—	32	13	45	—
Assam	99 (160)	1,214 (2,251)	1,313 (2,411)	129 (205)	3.809	1.052	1.35	2.006	138	466	649	94
Bihar	126 (188)	1,856 (6,159)	1,982 (6,347)	1,340 (3,193)	4.914	1.63	1.84	3.354	226	1,105	1,331	1,640
Goa	5.1 (—)	16.36 (—)	21.46 (—)	16.38 (—)	5.413	1.85	2.68	3.292	10	11	21	20
Gujarat	98 (133)	1,263 (2,060)	1,361 (2,193)	2,194 (3,323)	8.053	3.00	3.36	3.963	287	1,383	1,670	3,174
Haryana	134 (193)	306 (451)	440 (644)	1,716 (2,551)	6.7	4.22	4.97	5.79	328	471	799	3,627

(Contd.)

TABLE 3.10 (Contd.)

(1)	(2)	(3)	(4)	(5)	(6)	(7)	(8)	(9)	(10)	(11)	(12)	(13)
Himachal Pradesh	105.48 (137)	308.7 (562)	414.18 (699)	319.62 (474)	3.385	1.74	2.16	3.069	129	197	326	359
J & K	—	—	—	—	—	—	—	—	—	—	—	—
Karnataka	447 (541)	1,973 (3,076)	2,420 (3,617)	1,713 (2,265)	6.25	2.11	2.87	2.66	1,020	1,520	2,540	1,662
Kerala	856 (1,226)	293 (400)	1,149 (1,626)	36 (50)	6.23	2.55	5.29	5.71	1,947	273	2,220	75
M.P.	103 (201)	4,395 (4,971)	4,498 (5,172)	2,522 (3,898)	5.594	1.32	1.41	2.995	210	2,115	2.325	2,758
Maharashtra	625 (—)	1,879 (—)	2,504 (—)	1,962 (—)	7.07 (—)	1.58 (—)	2.95 (—)	3.699 (—)	1,613 (—)	1,087 (—)	2,700 (—)	2,650 (—)
Manipur	11.89 (21.97)	44.81 (102)	56.7 (123.97)	11 (23)	6.5	1.45	2.51	3.25	28	24	52	13
Meghalaya	10.449 (13.385)	86.49 (182.91)	96.90 (196.30)	5.448 (11.78)	8.81	0.75	1.61	1.8	33	24	57	4
Mizoram	4.6 (5.9)	23.85 (32.11)	28.45 (38.01)	—	6.39	1.05	1.93	—	11	9	20	—
Nagaland	—	—	—	—	—	—	—	—	—	—	—	—
Orissa	189.25 (316.42)	1719.54 3821.01	1909.39 4137.43	209.06 (402.15)	3.925	0.50	0.85	1.845	272	317	589	141

Punjab	564 (880)	183.9 (333)	747.9 (1,213)	2.320 (3,620)	8.703	2.79	7.25	6.322	1,792	187	1,979	5,355
Rajasthan	32 (47)	2,215 (4,467)	2,247 (4,514)	2,610 (4,460)	5.31	2.78	2.82	4.107	62	2,248	2,310	3,913
Sikkim	—	—	—	—	—	—	—	—	—	—	—	—
Tamil Nadu	667 (880)	1,264 (2,166)	1,931 (3,046)	1,221 (1,937)	5.9	2.48	3.66	3.8	1,435	1,144	2,579	1,694
Tripura	—	—	—	—	—	—	—	—	18	55	73	—
U.P.	421.51 (654)	p555.78 (6,174)	3977.78 (6,828)	6369.6 (9,889)	6.127	2.14	2.56	3.901	943	2,776	3,719	9,069
West Bengal	366.6 (511)	2,835.8 (4,828)	3,202.4 (5,339)	150.7 (216)	7.239	2.07	2.66	5.716	969	2,147	3,116	314
Union Territories	51.63 (69.19)	17.957 31.051)	69.587 (100.24)	130.04 (151.98)	5.36	2.60	4.64	5.77	101	17	118	273.74
All India Total **	5,095 (6,411)	26,783 47,359)	31,878 (53,770)	28,310 (41,909)	6.46	1.78	2.56	3.91	11,361	18,445	29,811	40,413

— Data not available; ** Total of only those states/UTs for which data are available. Figures in brackets show total number of milch animals.

Source : Animal Husbandry and Veterinary Department, States/UTs, etc.

Notes and References

1. Aneja, R.P. (1980) : *'Breed & Butter Potential of Dairy Development'*, Kurukshetra, Vol. 28(20), pp. 6–13.
2. George, J. (1988) : *'Dairy Development Policy'*, Instruments and Implications for India, Oxford Agraman Studies, 17, pp. 62–75.
3. John, Patric (1966) : 'दूध अमृत है, इसे बर्बाद न करो'। बिहार समाचार, अप्रैल।
4. Gangawar, D.S. (2003) : *'Sudha Bihar ka Gaurav, Hindustan'*, 19th May, Managing Director Bihar State COMFED.
5. John, Patric (1970) : *'A Case for Setting-up Small Dairy Farm'*, The Searchlight, Patna
6. Planning Commission (1951–56) : Sixth Five Year Plan.
7. Guruswami, Mohan (2004) : *'A Study in contrast in Punjab and Bihar'*, The Hindu, Business Line, 27 Sept. 2004.
8. Davis, R.F. (1965) : *'Modern Dairy Cattle Management'*, Printice Hall of India, New Delhi.
9. Whyle, R.O. & M.L. Mathur (1968) : *'The Planning of Milk Production in India'*, Orient Longman, Bombay.
   WTO (1995) : *'The World Dairy Markets for Dairy Products'*, Nos. 95-3814, Geneva, The WTO Ministerial Conference.
   Amble, V.V. (1965) : *'Milk Production of Bovines in India and their Feed Availability'*, Indian Journal of Veterinary Science, Animal Husbandry, Vol. 35, Nos. 221-33.
10. Bhole, R.A. (1967) : *'Trends in Milk Production During 1956–61'*, Agricultural Situation in India 22(6), 663–74.
11. Jain, R.L. & Gupta (1980) : *'Organised Milk Marketing in India : Social & Economic Impact : A Study of DMS in NW Rajasthan'*, D.K. Publishers, New Delhi.
12. Planning Commission (1951–56) : First Five Year Plan.
13. Do : Second Five Year Plan.
14. Do : Third Five Year Plan.
15. Do : Fourth Five Year Plan.
16. Do : Fifth Five Year Plan.
17. Do : Sixth Five Year Plan.
18. Singh, R.K.P. (1994) : *'Dynamics of Diary Development in Bihar'*, Indian Diaryman, 46, 8, 94.
19. Planning Commission (1951–56) : Seventh Five Year Plan.
20. Planning Board (1987) : *'Report of the Task Force in Development of Animal Feeds & Fodders'* (1985–2000 AD), State Planning Board, Bihar, April, 1982.
21. Planning Commission : Eighth Five Year Plan.
22. Planning Commission : Ninth Five Year Plan.
23. Planning Commission : Tenth Five Year Plan.
24. Shukla, Shruti (1993) : *'Dairy Project a Smashing Success Story'*.
25. Bureau of Public Enterprises (1984) : *'Annual Report on State Public Undertakings'*, 1983–84, Dept. of Finance, Govt. of India, Patna.
26. Census (1991) : *'Population Figures at a Glance'*, Directorate of Census Operations, Bihar, 1992, p. 2.
27. Pankaj, Prabhat Kumar *et. al.* (2005) : *'Animal Husbandry and Dairying'*, Kurukshetra, December, 2005.

Dairy Co-operation, Education, Training and Dairy Technology Development

DAIRY CO-OPERATION IN INDIAN STATES

Co-operation in dairy sector occupies prominent place in India as well as in Bihar. Dairy co-operation has completely changed the scenario of dairy farming despite its complete failure in other sectors. Co-operation is an old human society. The great philosopher Aristotle has rightly stated, "*Man is a social animal*'. This basic statement will hold good as long civilization exists in this world. Truly, it is the basis of domestic and social life. If the man lives in isolation he can't lead a happy and contended life. In practical life he really needs company to live and work help and support in times of stress and strain. In fact, *the history of modern civilization is the history of co-operation*. Without co-operation the social and economic progress would have been impossible.[1] Consequently, there is a greater need for understanding co-operation which can be considered as the basic principle underlying human life.

According to modern view the driving force in a human being is the accomplishment of desired objectives defined by his motives, needs and aspirations.[2] Human needs and behaviour are the reflections of the biology, spirit and mind of human being which are further natured and shaped by society, ecology and process of development. These co-operative forces are biologically very important and vital.[3] (R.S. Jalal, 1996)

The co-operative thought which was propounded for the first time of first half of the 19th century in Britain and France and which was given the first organisational structure in 1944 in Rochadale, England—spread for and wide and penetrated into almost all industries—specially into those allowed with agriculture.

Co-operation is one of the economic miracles of the last century. Both developed and developing are economically interdependent. Each one of these countries needs the co-operation of other countries and the help of developed countries to initiate and intensify their process of industrialisation. Even today, the developed countries are dependent on other developed countries for the raw material and have been largely influenced by the extent of the market created by the under-developed countries for their capitals goods. That is why independent countries, however powerful they might be, are striving hard to establish constructive relationship with other countries.[4] (Sharda, V., 1986)

According to the well-known co-operator, E.R. Brown-Co-operation is the universal instrument of creation, Prof. J.S. Mill the eminent Economist in his book '*Principle of Political Económy* stated '*Co-operation is the noblest idea*. It transfer human life from a conflict of classes struggling for opposite interests, to a friendly rivalry in the pursuit of the common goods of all. In the words of A.E. Emerson 'co-operation has acted more powerfully towards development of man and the better competitive struggle for existence'. Noted French Economist Charles Gude has announced that competition is essentially a kind of warfare which means the triumph of the strong and the rain of the week. Thus, it is co-operation and not conflict which motivates and directs human life towards the pursuit of peace and prosperity.

It brings together people and nation, and facilitates peaceful co-existence."[5] (Sharda, V., 1986)

No doubt, co-operation is a primary stage of civilization. In ancient period, men lived together in group and protected their interest collectivity. In country like India, where the concept of co-operation is still reflected in the joint family system. Village community were almost entirely structured co-operative.[6] (B.S. Mathur)

The modern concept of co-operation is the result of the great Industrial Revolution of Great Britain. The term 'Co-operative' was for the first time used by Rochdace Pioneer' and co-operative movement started from the Great Britain spread all over the world.

Co-operation is a new ideology which developed as a reaction against capitalism.[7] It grew out of the circumstances of the Industrial Revolution. The modern phenomenon of co-operation is thus, an economic concept and is of a formal nature. It is an organisation of people where the production activity is conducted by assigning a secondary role of capital. The co-operative movement, therefore, offers a ray of hope to the economically weaker section to live in better conditions of life in the modern world.

According to Prof. V.L. Mehta, a leading authority of Indian co-operatives, "a co-operation is a vast movement which promotes voluntary association of individuals having common needs who combine towards the achievement of common needs." According to Huges, "in its broadest sense, co-operative may be defined simply as a voluntary association in a joint undertaking for a mutual benefit." In the words of Prof. P.E. Smith, "Co-operation is an association belonging for service to themselves in which the risk of profit or loss is done by a variable price of goods and services rather than profit and capital."

P. Lambert has also defined, "a co-operative society is an enterprise formed and directed by an association of users, applying within itself the rules of democracy and directly intended to sense both its own members and community as a whole."

In 1950 Sir Edward Mecalagan defined, "Co-operation as the theory which maintains that an isolated and powerless man can, by association with others and by moral development and mutual support, obtain to that extent the material advantage available to the wealthy and powerful person and thereby develop himself to the fullest extent of his natural ability."[8]

DAIRY CO-OPERATIVES

The co-operative thought which was propounded for the first time in the first half of the 19th century in Great Britain and France and which was given the first organisational structure in 1944, in Rochdale. England, spread for and wide and penetrated into almost all industries—specially into those allied with agriculture.

In Denmark, the co-operative movement entered the dairy Industry for the first time, in 1882, when the Danish dairy farmers founded their first co-operative dairy. Only 20 years later, Denmark had about 1000 co-operative dairy societies each running a dairy plant. In 1963, the number of the co-operative plants rose to 1100 which handled 90% of the total milk delivered. These co-operative dairies began to hold the central position in the movement. These dairy co-operative have organised themselves in country-wide co-operative butter association which are reported to handle 67% of the butter export, 60% of the cattle feed and 40% of cattle export. In short, co-operative movement has been of incalculable importance to the whole economic and cultural development of Denmark.[6]

"Now-a-days, dairy countries like Newzealand, Australia and Holland, etc. have opted almost entirely the co-operative system in their dairy industries.

In most parts of Holland, the co-operative movement is traditionally very strong in the dairy sector. For many decades 85% of all milk produced has been going to co-operative dairy factories for processing.

"After World War-II, some 10,000 dairy farmers in the area surrounding Holland's big cities joined together and formed 'Co-operative Milk Centre (CMC)' which processed 10% of the total milk production in that country."

In U.S.A., a considerable section of the dairy industry is organised along co-operative lines. In Norway, practically all dairies are owned by producer co-operative organisation.

According to Shah, New Zealand, Sweden, Germany and U.S. where dairying is well advanced, more than three-fourth of the, milk produced is sold through producers' co-operative.[9]

In India, the picture is just opposite. In the year 1950 our co-operative dairies handled only 17% of the milk consumed in towns. The main reason of this poor handling was that among all democratic countries, India is the only country where majority of the dairy plants are in the public sector, although our Government as well as the Planning Commission have laid down that dairying should be developed along co-operative lines in preference to all other systems.

In fact, the field of Indian dairying is sufficiently fertile, for co-operative principles to grow and flourish. The small scale milk production on hundreds of scattered holdings situated for from marketing centres, absence of good roads, the presence of a large number of middlemen in the long chain of marketing, low purchase price-all these factors cry for co-operative movement in the dairy field. Under present system of exploitation by middlemen, a farmer does not develop the financial interest in the whole business of production, processing and marketing of milk twice a day and every day of the year.

In India, the co-operative movement entered the dairy field in the later half of the 19th century with the formation of cattle breeding co-operative societies. The societies maintained pedigree bulls and pasture lands and advanced loans for the purchase of good milk animals. In 1943-44, is all, 824 societies were operating. Their number, however, dropped down after independence when the government took the responsibility of cattle development activities in the whole of the country. In the first half of the 20th century, three types of co-operative organisations came up in the dairy sector. They were:

(i) Milk Producers' Co-operative Societies.
(ii) Milk Distributors' Co-operative Societies, and
(iii) Milk Consumer Co-operative Societies.

These co-operatives are known to render valuable service to the citizen during war periods. They checked rocketing prices and maintained the quality of milk and milk products.

In 1946, the Sarayya Committee recommended that all towns and cities with a population of more than 30,000 should be taken up for development work. The Committee also recommended to organise co-operative milk producers' societies around them within a radius of 30 miles. So far the pasteurisation and distribution processes of milk are concerned, the committee recommended to form Consumer's Milk Distribution Societies. The recommendation, however, was rejected in the 15th conference of co-operative Registrars. The Co-operative Registrars held that Distribution societies should be organised in place where no milk union existed.[10]

After independence the number of milk Co-operatives has increased. In the month of June, 1964, there were 126 all types of milk unions in the company.[11] The number of Primary Milk Producers' Societies was, 1300 in 1964-65 with 570,000 members. Besides, there were 200 Ghee Societies with 5,000 additional members. The value of milk and ghee sold by them in that year was Rs. 1,44,530,000 of total 72 dairy plants operating in the country in 1966, 19 were functioning in the co-operative sector, 49 in public sector and 4 in private sector. All the dairy plants in private sector were milk product factories. That year, 55 more dairy plants were under implementation and 21 were approved but not taken up. The share of co-operation was about 26% of the total number of plants. The percentage of the total daily throughput of all the plants in co-operative sector was estimated to be 24.6%. The details of the share is given in the following Table 5.1.

Table 5.1
Share of 3 Sectors in Dairy Industry in 1966

Sector	*Total*	*Average per plant*	*Total*	*Average per plant*	*% of utilization of targeted capacity*
Co-operative	615,000	32,400	396,243	20,850	64.43%
Public	1,748,000	38,000	1,077,462 .	23,400	61.64%
Private	N.A.	N.A.	137,588	34,400	N.A.

In the month of June 1967, there were about 135 milk unions having 28,566 members and a total working capital of Rs. 927 lakhs. The total number of milk supply societies was 8,911 with 6,60,021 members and a sum of Rs. 676 lakhs as working capital.

Thus, co-operation/co-operative organisation contains following *characteristics* and significance :

1. A co-operation organisation even in dairy farming is a voluntary association of persons and not an personal grouping of capital like joint stock company.
2. A person is free to join co-operative society and resign from his membership.
3. It is an entirely economic enterprise and not an ethical or charitable organisation.
4. Society is run by members at their own expense. It can earn profit or loss.
5. Its primary function is to serve the community as a whole and not just gain the profit.
6. Its basic nature is equality. In the ground of caste, religion, race, creed, political qualification no discrimination among the members is permissible.
7. All members has equal rights and duties. Its 'one member one vote' is a perfect example of equality.
8. Democratic management is its important feature.
9. All members of co-operative society enjoy equal right of voting, irrespective of numbers of shares held by them.
10. Capital does not get any special treatment over human beings.
11. The co-operatives have gone a long way in the expansion in a variety of activities in different sectors.

OBJECTIVES OF DAIRY CO-OPERATIVES

The objectives of dairy co-operatives all over India and especially in Bihar are always in conformity with the co-operative principles such as :[12]

1. To encourage the growing of fodder crops by its members or its affiliated societies members.

2. To purchase, pasteurised and sell milk and install the necessary plants for the same.
3. To render veterinary service and to provide medicines for cattle.
4. To buy animals on behalf of members or members of the affiliated societies.
5. To make arrangements for the disposal of the milk and milk products of its members or the members of its affiliated societies on a commission basis or to purchase out right and sell the same.
6. To settle all matters of common interest to the affiliated societies and to further their interest.
7. To control and supervise affiliated societies.
8. To own a herd of cattle for producing milk for sale with a view to meet any deficiency in supply on the part of members or members of its affiliated society.
9. To purchase and distribute to its members or members of the societies cattle feed and such other things as they may require.

In India in general co-operation in dairy sector is being encouraged and some extent in all branches of economic life. Like earlier plans Tenth Five Year Plan has also given much stress on the development of dairy industry on co-operative line. But the emphasis on dairy co-operatives as recommended in the Tenth Plan has two special characteristics; *First*, the plan envisages to modify the pattern of organisation of primary milk producers co-operatives. It is suggested that each primary society should be so enlarged that it may be able to have as an average a minimum daily milk collection of about 150,000 litres. Unless the collection is enlarged, the society can not invest in milk chilling equipment. *Second*, the Tenth Plan suggests to work towards a progress co-operativisation of Government milk plants, so that the entire chain of operations from milk, collection to transport pasteurisation and distribution get integrated.

In the words of Dr. (Mrs.) Anita Patel the Chairman of NDDB, "Co-operatives which continue to be the dominant player therefore have a special responsibility. They cannot afford to lose any time in removing the shackles of the past of and where necessary, adopting a professional approach as that is the only route to survival."[13] (Amrita Patel, Indian *Dairyman*, 2004, Vol. 56, No. 10).

PERFORMANCE OF MILK CO-OPERATIVES IN VARIOUS STATES

Dairying in India like all agricultural enterprises, rests with the individuals as small size units. Most of the farmers belonging to the small or marginal category, owning two or three cattle less than 2 hectares of land are attached with it. These farmers are scattered all over the country and are valuable to a variety of forces peculiar to the dairy industry (*Mescarenhas*, 1988). India accounts for about 1/6th of the cattle and 3/2nd of the buffalo population of the world. But, despite this large population the per cattle milk production is one of the lowest in the world. There is a ten-fold difference in declared milk output per cow from approximately 500 kg. per annum in India to over 5000 kg. in North America and some European countries[14] (*Dairy India*, 1992).

According to the Ex-Vice-President of India, Sri Bhairo Singh Shekhawat in his inaugural address to the National Dairy Development Board (NDDB), "The growth and development of Indian dairy industry since Independence is a success story that makes one very proud. The year of celebration of 50 years of our Independence also saw India achieving the landmark of becoming the highest milk producer in the world. Apart from tremendous growth in milk production, processing and value addition, there has also been quantum jump in dairy machinery, manufacturing, ancillary and packaging industries. From being primary import development, the dairy sector has raw acquired the exporter status."[15]

The milk co-operatives in India and particularly in Bihar in the form of COMPFED have played very important role in the success achieved in dairy development.

Dairy industry is an industry that has integral links with the farmers. Dairy development is a shining example of how the integrated system approach can help small and marginal farmers in the rural areas and also urban consumers, benefiting both at the same time.

Indian model of dairy development has shown how small farmers can join together and acquire the strength of a mountain by organized efforts through adoption of new technologies and also techniques of modern management and international marketing.

No doubt, dairy sector should be considered as the best friend of farmers. The dairy industry not only generates employment but will also strengthen the economy of our farmers in their villages itself.

Today, when there is a growing concern for greater attention to our rural economy, the dairy sector offers big opportunity to transform our economy by bringing prosperity to the rural sector including Bihar. Livestock husbandry as allied agriculture activity of the farmer has all through been a part of way of life and social ethos is rural India. The supporting income from animal husbandry and dairying sector in farmer's cash insurance against any distress caused by the crop failures.

The task of poverty alleviation is the most formidable challenge. Our 26 crore people still live below the poverty line. The dairy sector provides us immense opportunities for evacuating poverty. We need to make dairy development on eve of core competence in the national programme of poverty reduction and rural prosperity.

According to former Vice-President Mr. B.S. Shekhawat, "For the success of dairy industry, we need to pay due attention to cattle development. The diminishing population of good cattle due to drought in many parts of the country, has been a cause of concern. We need to concentrate on development of our own ethnic cattle breeds Rathi and Tharpakar that will give high yields and also provide sturdy bullocks for agriculture operations. Our rich stock of cattle breed with backup support of an integrated system of comprehensive services including provision of paper feed and fodder banks and veterinary care would provide a solid base for optimal success of the dairy programme."[16] (*Indian Dairyman*, 2004, Vol. 56, No. 10)

The new globalised economic world order, no doubt, offers opportunities for increased export led growth. However, there is a need to protect and safeguard against the unhealthy competition from the rich and highly subsidised farmers of the developed world. The best safeguard course, lies in building intrinsic competitive strength in our own industry. We should ensure the expansion of the infrastructure facilities. We need big quantitative improvement, right from the primary production point to the marketing and with our outputs conforming to the best of international standards of quality and reliability. We should first overcome all the deficiencies that so far have been restricting the optimum growth of our dairy sector.[17]

According to him, the success of dairy industry implies not on just increased production and exports of the dairy products, the real success is when every child gets at least a glass of milk every day, when no child suffers from malnutrition and no farmers suffer from distress of crop failure.

NATIONAL DAIRY DEVELOPMENT BOARD (NDDB) AND DAIRY CO-OPERATION

National Dairy Development Board (NDDB) is the apex body and headquarter of all dairy co-operatives in India. The NDDB was constituted in 1966 after the then Prime Minister of India Mr. Lal Bahadur Shastri who visited *'Anand'* (Gujarat) and spent the night in a village having population around 300 as a personal guest of a farmer. The PM went to that village, ate with the farmer and his family members. He began to walk around the village including village co-operative society. He saw men, women and children standing in a queue carrying milk in their hands and on their heads in clean utensils. All waiting patiently for their turn to poor milk. The PM saw that these may well be the first few steps in rural sanitation. The PM studied everything, low milk was collected and low farmers stand in a queue how the payment was made properly and how a truck came in time to pick up the milk collected by the society to transport milk to Anand. The PM saw milk from Brahmins, Muslims, Harijans going into the same *case*. He saw the caste, creed and religion submerging in that cane of milk"[18] (Dr. V. Kurian, *Indian Dairyman*). He appointed Dr. V. Kurien, the so called father of 'White Revolution', as the chairman of NDDC with the headquarter at Anand (Gujarat). The success of Anand was only due to dairy co-operatives. Dr. Kurien stated, "the success of Anand is that this is a dairy owned by farmers.

There is a dairy managed by elected representatives of the farmers. And since, it is their dairy it is bound to be so."

After Independence milk revolution in the form of White Revolution in India began in that fashion when Sardar Patel told a group of villagers from the Anand area in the Kheda district of Gujarat. Why do not you form your own co-operatives to market your milk ? That was in the early years of independence and the villagers were complaining to him about the exploitation they were subjected to a big private dairy in Anand town."[19] (*The Hindustan Times*, Sunday Magazine, May, 1985)

The experiment that a few villagers began then of co-operative procurement of milk at a central point led to co-operative marketing. Soon all three aspects—production, procurement and marketing of milk were strung together in the co-operative fold and Anand's experiment became a shining example for the rest of the country.

The Anand pattern envisages a three-tier model. At the lowest tier the producers at the village level form village milk co-operative societies managed by elected representatives. Any person who is a resident of the village and owns a milk cow or buffalo can become a member and avail of the facilities of milk collection, fat tests, daily payment, balanced cattle feed, fodder seeds, facilities for artificial and first-aid/veterinary service.

All village producers' co-operative societies federate into a district co-operative milk producers' union, which is managed by a board of directors, the majority of whom are elected by the chairman of the village co-operatives. The district union organises milk collection from villages twice a day, owns the processing plant, tests milk for fat contents, and provides a host of supportive services.

The district unions in a state form of federation at the state level, which is supposed to develop and implement marketing policies, establish a strong distribution network and norms for quality control, provide production schedules for member unions and participate in the national milk grid.

INSPIRATION BEHIND ADOPTION OF ANAND MODEL

Ex-Prime Minister Lal Bahadur Shastri during his official tour of Gujarat in 1964, expressed his desire to visit Anand.

Dr. Kurien took him on a conducted tour, after which the prime minister said be wished to stay incognito in a village for one night to see for himself what normal working conditions were like. The next morning a very impressed Shastri called upon Kurien and asked him to replicate the same model all over the country. The reasons the P.M. cited for his decision were that although the basic imperative for a successful milk co-operative was the fact that animals must be very healthy, the buffaloes of Anand were small, giving only two litres of milk, while in Punjab and Uttar Pradesh buffaloes give 5 to 10 litres, there was no advantage in having these animals but nevertheless Anand collected 2,00,000 litres of milk per day from a district. The soil was not as fertile as that in Uttar Pradesh and Punjab and the crop production was also just average. But still the experiment was a success. The critical variables lay elsewhere. It lays in the initiative, hope and faith of the simple villagers, faith in their ability to achieve the impossible. We may start with milk, but this tenacity of purpose is contagious, it is bound to spread to other vocations. Such an organisation structure conducive to their needs can act as a catalyst of socio-economic change. That was the rational behind setting up the National Dairy Development Board in 1965, to replicate not Anand, but the spirit of self-reliance and hope witnessed at Anand.

The effect of this pattern can be felt today in so far as by January, 2006, nearly 400 lakh litres of milk is being collected every day of which 167 lakh litres is sold as liquid milk in 186 metro and class I cities. Per capita availability of milk and milk products has gone up from 101 grammes per day.

The Anand experiment was human development in the correct sense of the word. An ordinary primary society member not only becomes economically more viable, but socially more responsible. In a society there are certain things he can do and which he cannot do, for instance he has to stand in a *quell*, he has to cast his vote every year, if he becomes a member of the managing committee he has certain obligations towards society. His responsibilities are so vast, that a small man becomes a much bigger man because of them.

According to the founding father of NDDB Dr. V. Kurien, the NDDB was given the task of implementing dairy development replicating Anand throughout India. It was at this time, when Dairy Board was formed, that we became aware of another threat—a very serious threat. Europe had a surplus of milk, and mountain of butter. But fortunately for us, that Chinese do not drink milk. So the destination for the surplus milk powder and fat was India. We realized that some kind of gentlemen from Europe would visit Delhi and would offer those milk powder and butter as gifts to India. We knew that this would land in India and would be sold free particularly in the constituencies of the ministers and politicians close to him. We knew that free availability of large quantity of milk powder would be disastrous. It would depress the prices of milk product produced by our farmers. The poor would become poorer. So, at Dairy Board, we thought of countering this threat.

The present chairman of NDDB Dr. (Miss) Amrita Paul has opined, "While we are looking forward at the scenario that will prevail over the next two decades it is perhaps appropriate that we also look back and recognize the enormous progress that has taken place in dairying since the 70's. We must acknowledge the invaluable contribution made by Dr. Kurien supported by able and dedicated colleagues overtime, in achieving the result through the operation flood programme in dairy development. The dairy industry has been fortunate that successive Prime Ministers also supported the programme and their governments put in place the necessary policies and ensured these were implemented."[20]

Today, we see enormous change in the competitive environment. The number of players engaged in the business of processing of milk has expanded manifold and this has thrown up both challenges and opportunities for all the responsible players. To continue this trend we should be producing more than 130 million tonnes of milk by 2015. It is desirable that as the milk production grows the share of the organised sector increases. At the same time, we must recognise that processing through modern dairy plants results in milk and milk product being supplied at a higher cost than the traditional system because of the better practice required to be followed in procurement, processing and marketing. The shift to modern milk processing consumer's purchasing power and how effective we are in educating consumer on the benefits of hygienic high quality milk. And this in turn will be possible only when government is able to enforce the existing food laws to ensure quality hygiene.

The essence of the challenge we face is to make certain that our dairy industry continues to keep its purpose constant, and it also continues to contribute significantly to the incomes and quality of the lives of the tens of millions of our fellow citizens who depend on it. If we are to meet this challenge, we can only do so as an industry. At the same time there is a genuine need to modernise our industry. And this requires linking producers directly with markets through professionally managed institutions.

DAIRY TECHNOLOGY

Dairy farmers have achieved quantum jump in raising their living standards through the use of dairy technologies the world over. The countries, which dominate the global scene are the ones, which happen to be technology leaders. Technology has been a major factor for bringing changes in agriculture, dairy, communication and other fields. The world is being reshaped by technologies and telecommunications speed up communications. (Chakravarty and R. Chand, *Indian Dairyman*, 2000)

Therefore, day-by-day the world is shrinking into a global village. 'Technology' comes from a Greek word 'techno' (the skill), *'loges'* means *knowledge. Technology* means the knowledge of how something is made. In the word of Nawaz Sharif, 'Technology' is A *'game'* for the rich, A *'dream'* for the poor, A *'key'* for the wise (1983, Management of Technology Transfer and Development APCTT, Bangalore).[21]

Technology is a tool or technique, product is process, physical equipment or method of doing or marketing (Goldring, 1976). Technology involves the application of science and knowledge to practical use, enabling man to live more comfortably and security (Hoda, 1979). Technology is a design for instrumental action that reduces the uncertainty in the cause effect relationships involved in achieving the desired got come (Rogers, 1982). Technology is systematic knowledge and action, usually of industrial processes, applicable to any recurrent activity (Mc Graw Hill, Encyclopaedia of Science and

Technology, 1982). Technology can be defined as the *'know-how'* to translate concepts into goods and services for the satisfaction of end users. *Kotler* defined, "A service in any activity that one party can offer to another that is essentially *infrangible* and does not result in the ownership of anything."

All the outcome of analysis of date cannot be said to be 'technology'. "The technology chain indicates how the findings of research projects may (or may not) be transformed into transferable technology. The various steps of the chain are : Findings, Feasibility, Testing, Introduction, Adoption, Diffusion, Impact, Changing people's customs is an even more delicate responsibility than surgery when a surgeon takes up his instrument he assumes the responsibility for a human life whenever with the technological change an administer seeks to alter a people's way of life, he is dealing not with one individual, but with the well-being and happiness so many generations of men and women." (Spicer, 1952).

NEEDS OF DAIRY TECHNOLOGY

As a method of increase the capacity of milch cows and buffaloes to produce more milk by judicious application of following recognised techniques of dairying are needed such as :

(i) *Breeding* : Mating animals of desired characters and selecting male and female offsprings.

(ii) *Weeding* : Removing uneconomic, unhealthy and unwanted ones.

(iii) *Feeding* : In proportion to body requirements and production of animals.

(iv) *Heading* : Day-to-day care and management.

(v) *Exposure* : Temperature, relative humidity and precipitation.

To increase the productivity of cattle considerable quantum of work has been done by the scientists in past two decades. With the available technical 'know-how' the scientists, Central Government and various State Governments have sponsored several livestock projects in the country, such as key village scheme, IRDP, Artificial Insemination, progeny testing scheme, etc.[22] (Jagdish Prasad, 1977)

As per the estimates furnished by F.A.O. during 1978, the world average milk yield per cow was about 1964 kg. and the estimated milk yield per cow in India during the corresponding period was 486 kg. which was one of the lowest in the world.

TABLE 5.2
State-wise Milk Production in India, 1971-72 to 1997-98

(Lakh MT)

State	*1971-72*	*1981-82*	*1991-92*	*1995-96*	*1996-97*	*1997-98*
(1)	*(2)*	*(3)*	*(4)*	*(5)*	*(6)*	*(7)*
Andhra Pradesh	11.3	24.2	30.57	42.61	44.70	44.75
Bihar	17.5	20.4	31.59	33.21	33.99	34.26
Gujarat	17.9	22.4	36.60	46.08	48.31	49.13
Haryana	15.1	22.7	34.58	40.55	42.04	490.82
Himachal Pradesh	2.7	3.4	4.92	6.76	6.98	7.14
J & K	2.3	2.6	7.47	8.62	9.00	9.38
Karnataka	7.6	12.0	24.90	31.84	34.60	39.70
Kerala	2.8	9.5	17.90	21.92	22.58	23.48
Madhya Pradesh	11.7	23.9	47.90	51.25	52.24	53.78
Maharashtra	11.9	17.7	39.19	49.91	51.27	51.93
Orissa	3.4	3.2	5.06	6.48	6.87	6.70
Punjab	21.4	34.9	53.63	64.24	67.55	71.65
Rajasthan	25.4	33.0	33.63	54.49	58.73	55.00
Tamil Nadu	9.3	18.4	34.22	37.91	39.76	40.00
Uttar Pradesh	43.0	59.5	101.71	118.78	123.87	129.34
West Bengal	4.9	17.8	29.68	3.41	33.76	34.15
All India Total	225.0	343.0	558.39	661.87	690.66	706.23

Source : CMIE, Various issues.

TABLE 5.3
Percentage of State-wise Share of Indian Milk Production

State	*1971-72*	*1981-82*	*1991-92*	*1995-96*	*1996-97*	*1997-98*
(1)	*(2)*	*(3)*	*(4)*	*(5)*	*(6)*	*(7)*
Andhra Pradesh	5.02	7.06	5.47	6.21	6.47	6.37
Assam	0.67	1.52	1.17	1.06	1.07	1.10
Bihar	7.78	5.95	5.66	5.02	4.92	4.88
Gujarat	7.96	6.53	6.55	6.96	6.99	6.99
Haryana	6.71	6.62	6.19	6.13	6.09	5.81
Himachal Pradesh	1.20	0.99	1.06	1.02	1.01	1.02
J & K	1.02	0.76	1.34	1.30	1.30	1.34
Karnataka	3.38	3.50	4.46	4.81	5.01	5.65
Kerala	1.24	2.77	34.20	3.31	3.27	3.34
Madhya Pradesh	5.20	6.97	8.58	7.74	7.56	7.66
Maharashtra	5.29	5.16	7.02	7.54	7.42	7.39
Orissa	1.51	0.93	0.91	0.98	0.99	0.95
Punjab	9.51	10.17	9.60	9.71	9.78	10.20
Rajasthan	11.29	9.62	7.99	8.23	8.50	7.83
Tamil Nadu	4.13	5.36	6.13	5.73	5.76	5.69
Uttar Pradesh	19.11	17.35	18.21	18.13	17.95	17.94
West Bengal	2.18	5.19	5.31	5.05	4.89	4.86
Others	6.80	3.56	1.14	1.03	1.01	0.50

Source : Basic Animal Husbandry Statistics, 1999, Department of Animal Husbandry and Dairying, Ministry of Agriculture, Govt. of India.

TABLE 5.4

State-wise Number of Milch Animals in India, 1992

(*Thousand Numbers*)

State	*Cattle (Cross Bred)*			*Cattle (Indigenous)*			*Buffaloes*			*%age*
	In-Milk	*Dry*	*Total*	*In-Milk*	*Dry*	*Total*	*In-Milk*	*Dry*	*Total*	
(1)	*(2)*	*(3)*	*(4)*	*(5)*	*(6)*	*(7)*	*(8)*	*(9)*	*(10)*	*(11)*
Andhra Pradesh	153	52	205	1269	902	2171	3192	1270	4462	6.97
Bihar	35	24	59	1767	3187	4954	948	1475	2423	7.58
Gujarat	95	26	121	1220	658	1878	2085	898	2983	5.8
Haryana	100	38	138	338	139	477	1496	511	2007	2.67
Himachal Pradesh	84	29	113	301	235	536	300	136	436	1.11
Jammu & Kashmir	167	49	216	417	264	681	253	127	380	1.30
Karnataka	184	74	258	1698	1698	3396	1220	985	2205	5.97
Kerala	583	259	842	467	292	759	73	32	105	1.74
Madhya Pradesh	59	28	87	4145	4284	8429	2088	1360	3448	12.20
Maharashtra	549	278	827	1999	2443	2442	1799	1141	2940	8.37
Orissa	153	85	238	1691	1986	3677	206	182	388	4.38
Punjab	456	186	642	262	124	386	2274	894	3168	4.28
Rajasthan	28	13	41	2100	2089	4189	2429	1340	3769	8.06
Tamil Nadu	506	219	725	1420	808	2228	979	441	1420	4.46
Uttar Pradesh	367	218	585	3356	2412	5768	6024	3388	9412	16.08
West Bengal	290	107	397	2943	1716	4659	162	59	214	5.37

Source: GOI, 1998.

TABLE 5.5

Growth of Village Milk Producers' Co-operative and Procurement under Operation Flood (1970 to 1990)

Year	*Member VMPCS*	*Annual Milk Producers ('000)*	*Procurement (Million Tonnes)*
OF-I			
1970-71	1588	278	0.1898
1971-72	1811	327	0.2372
1972-73	2200	361	0.2774
1973-74	2598	394	0.2226
1974-75	2966	445	0.3175
1975-76	4533	562	0.4197
1976-77	7681	723	0.5657
1977-78	9306	943	0.6205
1978-79	10099	1213	0.7336
1979-80	11436	1475	0.8614
1980-81	13270	1747	0.9844
OF-II			
1981-82	18422	2125	1.0147
1982-83	23496	2620	1.6133
1983-84	28614	3116	1.9016
1984-85	34523	3632	2.1097
OF-III			
1985-86	42692	4484	2.8762
1986-87	49077	5097	2.8652
1987-88	54525	5666	2.8105
1988-89	58883	6250	2.9090
1989-90	60825	7003	3.5821

TABLE 5.6
Percentage of Cross-breeds to Total Cattle

State	*Cross-breeds*	*Indigenous*	*Total*	*(Per cent) Cross-breeds*
Andhra Pradesh	173.00	13047.00	13220.00	1.31
Assam	145.00	660.50	67.50	2.15
Bihar	151.00	16062.00	1521.30	0.93
Gujarat	53.00	6941.00	6994.00	0.76
Haryana	266.00	2076.00	2342.00	11.36
Himachal Pradesh	124.00	2050.00	2174.00	5.70
J & K	163.00	2162.00	232.50	7.01
Karnataka	538.00	10762.00	11300.00	4.76
Kerala	145.30	1644.00	3097.00	46.92
Madhya Pradesh	66.00	270.51	27117.00	0.24
Maharashtra	492.00	15670.00	16162.00	3.04
Manipur	62.00	685.00	747.00	8.30
Meghalaya	18.00	532.00	550.00	3.27
Nagaland	31.00	120.00	151.00	20.53
Orissa	227.00	12703.00	12930.00	1.76
Punjab	N.A.	N.A.	N.A.	N.A.
Rajasthan	39.00	13466.00	1350.50	0.29
Sikkim	33.00	140.00	173.00	19.08
Tamil Nadu	885.00	9480.00	1036.50	8.54
Tripura	35.00	6.44	680.00	5.29
Uttar Pradesh	3252.00	22900.00	261.52	12.43
West Bengal	554.00	15103.00	15657.00	3.54
Union Territories				
A & N Islands	2.00	34.00	36.00	5.56
Arunachal Pradesh	0.00	168.00	168.00	0.00
Chandigarh	2.00	4.00	6.00	33.33
D & N Haveli	1.00	44.00	45.00	2.22
Delhi	4.00	48.00	52.00	7.69
Goa, Daman & Diu	3.00	129.00	132.00	2.27
Lakshdweep	1.00	2.00	3.00	33.33
Mizoram	3.00	46.00	49.00	6.12
Pondicherry	25.00	68.00	93.00	26.88
Total	8802.00	180386.00	189188.00	4.65

Amongst the states in the country, Uttar Pradesh stands first in milk production with an estimated production of 12,934 thousand tonnes in 1997-98, followed by Punjab with 7,165 thousand tonnes, Rajasthan 5,500 thousand tonnes and Madhya Pradesh 5,378 thousand tonnes. During 1981-82 the top five milk producing states were U.P. (17.35%), Punjab (10.17%), Rajasthan (9.62%), Andhra Pradesh (7.06%) and M.P. (6.97%), according for more than half of the total milk production. However, in 1997-98, Andhra lost its position among the top five milk producing states. In 1997-98, top 5 milk producing states were U.P. (18.41%), Punjab (10.20%), Rajasthan (7.83%), Madhya Pradesh (7.66%) and Maharashtra (7.39%). These states account for about 50% of milk produced in the country. The share of Gujarat, Karnataka, Kerala, M.P., Maharashtra, Orissa, Punjab, Tamil Nadu and Uttar Pradesh in total Milk production increased between 1981-82 and 1997-98. In contrast the share of Andhra, Bihar, Haryana, Rajasthan and West Bengal declined during this period. In general the eastern states have a very low milk production.

The share of Gujarat, Haryana, Punjab, Tamil Nadu and Uttar Pradesh in total milk production is substantially higher than their share in breedable bovine population on the other hand, Andhra Pradesh, Bihar, M.P., Maharashtra, Orissa, Rajasthan and West Bengal have lower production of milk relative to their population.

LIFE CYCLE OF DAIRY TECHNOLOGY

Most technologies related to dairying follow a life cycle ranging from birth to decline in their use as demonstrative. The time period between these stages however, varies to a great extent. This is due to the changing needs. "Technology has no inherent value in itself and no value to society until it is applied for the purpose for which it was created. The central question than is what kind of delivery system and policies are needed to bring positive impacts to extended clients and to reduce the negative impacts on non-intended audiences." (Leagans, 1980).[23]

For selecting the appropriate technology and its effective dissemination in the dairy farming in Bihar, the following steps are imperative :

(a) Identification of facts related to problem.
(b) Selection of most likely reasons related to problems.
(c) Determine the knowledge skill, attitude and aspirations needed to alleviate the problem.
(d) Decision on educational objectives and experiences.
(e) Clientele.
(f) Teaching methods.
(g) Period.

ELEMENTS IN THE TECHNOLOGY TRANSFER

In the process of technology transfer, following three elements are very important particularly in Bihar and other states like U.P. and M.P. : (i) Technology, (ii) Extension practitioners, (iii) End-users.

The extension practitioners act as a calalyst to brisk/hasten the process of technology transfer.

For extension practitioners following eleven competencies are more popular :

(i) Extension workers need to know and understand the extension education, its objectives, organization and relationship to the institution.
(ii) Extension workers need to know and understand technical subject matter appropriate to their needs and the needs of people with when they work.
(iii) Extension workers need to know and understand the principles and process of programming and to have a high degree of proficiency to applying these concepts.
(iv) Extension workers need to know and understand the principles of learning and teaching and to have a high degree of proficiency in applying these principles.
(v) Extension workers need to understand and to have a high degree of proficiency in communication process.
(vi) Extension workers need knowledge about and understanding of the structure and dynamics of human society.

(vii) Extension workers need to understand human development process and to maintain a high degree of skill in human relation.

(viii) Extension workers need to understand the principles of management and to attain a high degree of proficiency is applying these principles.

(ix) Extension workers need to be informed about current issues and problems confronting the people and be proficient in discussing them in an objective and informative manner with groups.

(x) Extension workers need to know and understand the principles and administration and supervision.

(xi) Extension workers need to know and understand and be proficient in applying are principles and techniques of evaluation.

ADOPTION OF DAIRY TECHNOLOGY

For a technology to reach its potential in bringing about a positive impact, its adoption is a prerequisite. Adoption is not a phenomenon, which just happens by chance. It is indeed, a process comprising five important stages which are :[24]

1. Awareness	:	First hear
2. Interest	:	Seek Information
3. Evaluation	:	Weigh up Pros & Cons.
4. Trial	:	Test the Technology.
5. Adoption	:	Apply the Technology.

Roger, E.M. (1983) explained that just as the attributes of the technology are vital, so are the categories of the adapters. These range from innovators to laggards. Studies have indicated that innovators are about 2.5 per cent of the distribution followed by 13.5 percent of adapters who take up the new technologies. They are followed by 34 percent known as early majority followed by an equal percentage of adaptors who take up the technology at a much later time the take majority who are finally followed by 16.0% of laggards who are the last category.

The best approach for better results was to work with farmers very closely with dedication and by narrowing down the turf of psychological contract between farmers, extension scientists and basic dairy scientists by increasing reasons for farming technology. The extension education can help to overcome dairy farmers' barriers i.e. (a) knowledge, (b) motivation, (c) resources, (d) insights, (e) power, (f) attitude, and (g) aspirations.

EFFECTS OF DAIRY TECHNOLOGY

The consequences are the changes that occur to an individual or to a society as a result of the adoption or rejection of a technology. In addition to the positive effects for fulfilment of human needs, technologies are sometimes associated with negative effect.

In the last changing values and jerky environment the technology transfer assumes greater role and significance. The care should be taken that the technologies, which are developed, should not bounce back as technologies are means for survival and for the progress of the development world. The technology is a tool box for dynamic change. The efforts should be made to increase the positive impacts and reduce the negative impacts. There is burning need to develop a mechanism for continuous monitoring of consequences of the adoption of technology. (R. Chakravarty & R. Chand, *Indian Diaryman*, 2004).

Need	*Positive Effect*	*Negative Effect*
Milk	Enhanced Milk productivity	Contamination, synthetic milk adulteration
Animal Health	Reduction in mortality	Side effects of medication shortage of fodder and feed
Extension	Dissemination of knowledge	Brain washing by mass media

PROCESS TECHNIQUES

Milk is the basic ingredient used in the preparation of Indian milk products. Consistency in the quality of milk is assured by agreements with milk suppliers, based on specifications covering fat, temperatures and processing parameters, aroma, flavour and microbial standards. Besides milk, other raw materials like sugar, spices, flour can introduce micro-organism, toxins and chemicals, which may after the safe status, wholesomeness and shelf life of product.

In manufacture of some products, buffalo milk is used exclusively while others are better made from cow milk. Whatever the type of milk used, it must have low bacterial counts, no off flavours and no enzymatic breakdown of fat or protein. To check the growth of microbial flora in milk, it should be cooled at 4°C immediately after milking and held at this temperature unit processed.[25]

Cream, butter, ghee and other milk concentrates used in the preparation of indigenous milk products, may also carry micro-organisms. For micro-biological safety of the end-product, these ingredients should also confirm to sanitary standards similar to milk. Concentrated fat products have vulnerable to rancidity problems unless stored under refrigeration.

All ingredients other than milk and milk derived products must also conform to stringent quality standards to remove chances of micro-biological contamination of milk products being manufactured.

WATER ACTIVITY

Water activity, plays a crucial role in predicting the shelf-life of foods and bringing in it the desired extension. All micro-biological activity requires water for growth in addition to nutrients and presence or lack of oxygen or air. Therefore, food preservation techniques are used not only for manipulation of available by replacing air with nitrogen and other inert gases in the food package. This is the basis of Modified Atmospheric Packaging (MAP) which is widely used for cheese and possibly could be used for paneer.

The relationship of water with food in package or in atmospheric storage is explained by water absorption. While absorption process occurs throughout the food surface and its interior 'sorption' is absorption that occurs only at the food environment interface. Sorption of water vapour to or from a food is a function of vapour pressure regulated by the moisture in the food.

The water sorption of dairy food is largely determined by their non-fat solids contents like lactose and protein as well as crystalline status and structural transformation going on in them.

Indian milk products have generally high water activity because of which they have short shelf-life. However, their water activity can be lowered by judicious use of glycerin and polyols, whatever feasible, reduced water activity can also be attained by adding salt, sugar, proteins and starch and corn syrup solids. Sugars bind water by their capacity to convert part of available water to water of crystallization. Similarly, proteins, starched, cornsyrups and sugers bind water via hydrogen bonding.

Burfi, peda, kalakand and other *khoa*-based products and sweetened condensed milk fall into intermediate moisture range, whereas chhena and *dahi*-based products have high moisture content. Processing treatment of high moistures foods are designed to extend their shelf-life significantly.

HURDLE TECHNOLOGY

While water activity and sorption behaviour are useful parameters, they do not solely control or decide the shelf-life or stability of a food. A better approach is to use them with reversal other parameters to make the product safe during desired shelf-life. The combined use of several parameters has been termed as *Hurdle Technology* or Combined Method Approach. This approach involves manipulating the p^H E_n (redox potential), A_w (water activity), solute type and concentration, giving heat treatments, incorporating chemical preservatives; using effective packaging techniques and maintaining chilled conditioned of storage.[26]

Through *Hurdle Technology*, a non-sterilized food can made safe for consumption as well as its taste kept fresh over an extended period of time to facilitate its marketing over a wide distribution network. Thus, the name *'Hurdle Technology'* refers to each of parameters listed above possess a hurdle to pathogenic and spoilage micro-organism. Several of these hurdle act in synergistic manner.

PROCESSING CONDITIONS

Heat treatment during and after processing of a food has a profound effect on its shelf-life. Different processing procedures could have antagonistic or synergistic effect on the food stability due to interaction of various ingredients or because of chemical reactions induced by the heat application.

In khoa manufacture, the intense heat treatment destroys the microbial population. The heat treatment also has a drawback—excessive heat can induce undesirable burnt flavour and insufficient heat can lead to improper flavour attributes. The culturing process also helps in checking the microbial growth.[27]

PACKAGING TECHNIQUES

As Indian milk products are microbian logically sensitive, they need hygienic handling during filling and packaging. The filling machines and the packaging materials must be so chosen as to achieve the best possible shelf life for these products. In most cases, the package should be non-flexible and firm enough to protect the product and withstand rigorous of transportation and distribution.

The design of the filling machine should be in consonance with the sanitary practices. It should be easy to dismantle the filler unless it is designed for emplace cleaning (CIP). It should have the arrangement for through cleaning and sterilization prior to the start of filling operation.

The main function of packaging is to protect the product from physical damage and environmental changes during transit. The packaging materials should also present deterioration of the product quality as a result of exposure to sunlight.

ASEPTIC TECHNOLOGY

Aseptic technology is the second biggest breakthrough in the field of food processing since pasteurization. Aseptic packaging offers benefits no other kind of packaging does. The reputed Institute of food technologist considers the aseptic processing and packaging of liquid foods to be the most important food science advancement in the 21st century.[28]

The pasteurised plastic pouch (PPP) has been a major packaging medium for liquid milk in developing countries.

PILLARS OF ASEPTIC TECHNOLOGY

The key to the introduction of an Aseptic Technology is most of the developing countries is low cost entry in terms of investment and the provision of an affordable costs in terms of operations :[29]

(i) The Processing System : (The Tetra Them Aseptic Prime Steriliser;
(ii) The Milling Machine; and
(iii) The Package : The Tetra Fimo (TFA) package.

THE PACKAGE

The package is a six layer aseptic packaging, shaped like the conventional pouch[30] :

(i) *Polythylene* : Provides moisture protection and print protection.
(ii) *Paper Board* : Provides printing for the package and studiousness.
(iii) *Polyethylene* : Bonding layers between paper and aluminium firmness.
(iv) *Aluminium Fool* : Key berrier layer provides barrier to oxygen, light, micro-organism, odours and flavours.
(v) *Polyethylene* : Binding layer between aluminium foil and inner layer.
(vi) *Polyethylene* : Seals the liquid inside, provides sealing of the package. The layer is compatible with the liquid to the packed.

The package has the possibility of a five colour flexographic printing which provides the possibility of having a consumer appealing package design. The use of paper within the package gives the package a better had feel and acts as a strong differentiation for the pouch.[31]

Thus, technological changes have been playing very significant part in white revolution in the states like Bihar. Productivity of milk has two components : (a) technical change, and (b) technical efficiency.

Efficiency is a very important factor of productivity growth especially in developing economies like India, where resources are scarce and opportunities for developing and adapting better technologies have lately started dwindling. It is still possible to increase production and productivity of milk in Bihar and even in India by improving efficiency. Efficiency of any dairy firm is measured in terms of its relative performance.[32]

In Production/cost relationship in dairying economists identify three types of inefficiencies : (a) Technical inefficiency, (b) Allocative inefficiency, and (c) Scale inefficiency.

DAIRY EDUCATION

As early as vedic period, veterinary science/dairy science was developed in India. Atharveda and Mahabharata (1500-500 BC) have references on dairy farming/dairy education, study of animal anatomy, horse management and health care aspects. The existence of veterinary hospitals has been well documented in Sanskrit scripts and edicts in different periods from the first millennium (Ranchandran, 1996, *"Vicissitudes in Veterinary Education Research & Extension : In Animal Production & Health in India"*, Technical Bulletin, No. 2, APAU, Hyderabad)

With the establishment of an army veterinary school in Pune in 1862, the modern veterinary education in India has made a beginning. As on today, there are 35 veterinary colleges producing approximately 2200 graduates per year. In parliament India felt the need to regulate veterinary practices and passed the Indian Veterinary Council Act, 1984.

The veterinary/dairy education is not only to seek or cultivate knowledge, but has to provide a veterinarian with skills and competence to lead society towards progress.

Dairy education is the science of breeding, feeding and tending domestic animals especially farm animals, white dairying/veterinary science deals with the study, prevention and treatment of diseases in animals.

BIOTECHNOLOGY AND DAIRYING

Biotechnology can be broadly defined as the technology by which we can produce useful products from raw materials with the help of living organism or other biological progress. This technology has been rightly called *biorevolution* as it has great potential to influence our life in a spectacular way. Since the last decade Biotechnology has sparked unprecedented interest in different areas of science, namely, medicine, industry, agriculture, animal husbandry, environment, etc. in order to improve the overall living standard of the human population.

Dairying is yet another area where biotechnology can make a great impact. In a developing country like India, improvement is the productivity of the milk animals in terms of high milk yield as well as production of clean milk and processing of the milk into dairy products of high quality can influence the economic status of our farmers, dairymen as well as Dairy Industry.

APPLICATION IS DAIRY PRODUCTION

The foundation stock of milk yielding animals such as cattle, buffaloes and goats in our country are genetically far inferior to those in western countries. Genetic improvement of these animals by traditional methods of breeding will take many decades. However, cross-breeding programme will yield quick results.

The recent developments in animal reproductive biology and genetic basis of traits have fostered new animals breeding technology. In future, animal breeding programme may be augmented to achieve desired changes with great speed and selection.

ANIMAL NUTRITION AND GROWTH PROMOTION

The use of biotechnology to improve animal feed and nutrition promises to improve the animal productivity.

Synthetic steroids and natural hormones are used widely to promote animal growth.

GENETICALLY MODIFIED FOOD AND INDIAN DAIRY

According to T.K. Wali the Principal Scientist of Animal Nutrition, "Man's quest for knowledge, coupled with the pressures to seek solution to the global problems of feeding billions of human beings and also the domesticated livestock, and of keeping them as much as possible free from diseases, led to break through in biotechnological research, culminating in the development of GMO, i.e. genetically modified organisms (plants, animals, micro-organisms), which are supposed to contain some additional superior traits compared to the conventional organisms." (*Indian Dairyman*, Vol. 55, No. 6, 2003, pp. 55-56)

Evidently, genetic engineering technologies and GMOs have the potential to significantly raise the levels of efficiency and productivity is plants and animal production and need to food requirement of the fast multiplying human race, where number will increase from the present level of 6 billion to 9 billion in the next few decades, while the agricultural land is diminishing through urbanisation and industrialisation and the deforestation is constituting a serious blow to fragile ecosystem.

"Government of India is still weghing the pros and cons of these issues, and is yet to take a decision, which should guide the future policy regarding the import or export of GM foods feeds, industrying issue of segregation. The problem that comes up among the users of GM ford is the availability of authentic long-term guarantees of financial viability, environmental safety, security against health risks, toxicity and above all assurance of consumption of GM foods would not cause gene transfer to the human and animal through got microflora." (*Indian Dairy*, Vol. 55, No. 6, 2003, p. 37).

COMMON PROPERTY RESOURCES AND RURAL DAIRY PRODUCTION

It should our prime objective to evolve new technologies in livestock rearing practices which will really fit with the existing situations of our dairy products. Thus, enabling them to near their stock intensively with reduced pressures on Common Property Resources (CPRs) thereby helping to maintain them in a productive and sustained conditions to sustain the livelihood of our millions and millions of rural poor and their stock on a sustainable basis.

Common Property Resources (CPR) may be defined as resources accessible to the whole community of a village to which no individual has exclusive property rights (Jodha, N.S., 2000, 'Wasteland Management in India, Myth, Motives & Mchanisms, *EPW*, Feb. 5).

CPR may be defined in several ways. It is worthwhile to define CPRs as finite resources which users have co-equal rights such as, village common grazing lands, community forests wahalanda, village ponds, community or village funeral sites, rivers, etc.

Broadly CPRs are of two types : (i) community managed CPRs, and (ii) open access CPRs. In community managed CPRs 'specific user groups', e.g. a village community, a specific tribal group, etc. manage and use the resources in a common and collective way where they obey the rules, restrictions or social sanctions prevailing within the user group regarding management, improvement and use of these resources and have the rights to exclude those who belong to another group.

In this case, CPR, may be owned by the 'user group' as a whole a may belong to another agency, e.g. village waterlands, belong to the Revenue Department of State through used and managed by village community. (Jodha, N.S., 1990 : *CPRs Contribution & Crisis, EPW*, June 30, 1990)

But contrary to this, in case, open access CPRs there does not exist any institutional or community management system to specify and identify the members, non-members among the resource users. These are owned by no one and belonging to everyone. These types of CPRs are really free to any users.

The CPRs not only provide fodder, grazing sites, leaves, to the animals of the poor, but also prove other valuable items of life sustenance such as fuel, branches, things, wood, fruits, etc. which they consume or sell to get cash money or utilise for some other purposes. On the other hand, their animals give milk, meat, skin, dung, etc. converting the resources they have been provided by CPRs.

Thus, CPRs play a vital role in rural economy in India. But unfortunately, in spite of their tremendous contributions they have received a little attention in the past, both from public and private sector.

The status of CPRs such as village forests, pastures, wasterland, watershed, drainages, water boodies (river, riverlets, ponds, estuaries, etc.) is also apt to modified, leading to serious socio-economic concern. (Sahoo & Mishra, 1994)

Recently their potentials are being recognised and felt. Consequently, a worldwide concern regarding their restoration, improvement and management is being observed keeping in view that CPRs should be managed and improved on a sustainable basis if they are to provide benefits to the animals and mankind on a long-term basis.

In India and even in Bihar animals are mainly in the hands of million & million of small and marginal farmers except a few commercial holdings. They have little or no access to cultivated land. So, they have to depend on CPRs to a great extant, which is otherwise not possible, with the changing situation of CPRs, they change their livestock composition and size to cope up with the situation. Accordingly, they tend to reduce their animal number.

The importance of CPRs is more in arid and semi-arid regions of the country where comparatively, the risk is high and the productivity is low.

But Indian policy of CPRs is quite defective, unscientific and irrelevant consequently richers become richer while poorer become poorer. All the government policies lacked proper CPR perspectives and participatory approach. (*Indian Dairyman*, Vol. 55, No. 10, 2003, pp. 63-66).

FUTURE OF GLOBAL DAIRY INDUSTRY AND OPPORTUNITIES FOR INDIA

Managing a vast market like India is a continuous process, intricate and complicated economic and political pressure build until a change the policy becomes inevitable.

The biggest issues affecting the future of milk policy will be the WTO agreements particularly AOA (Agreement on Agriculture), consumer health education and purchasing power. Besides, facing cost concern and competition *vis-a-vis* existing products, the dairy products, the dairy industry will undergo a major change.

Fuelled largely be new consumer demands based on scientific development that are supported by media exposure.

The global opportunities for the Indian dairy industry won't be in conventional areas like milk powder, butter, condensed or evaporated milk, etc. Dairy products are an integral part of the world's dairy diet.

Considering that commercial, technical, social, legal, economic and food safety factors affect food products across the globe, it is likely that they will influence the business of dairy products too.

The scope for growth in the dairy sector will depend upon the dairy industry's ability to service the market according to the need of its consumers. This will be based on key factors like convenience, quality, health and nutrition, entertainment and product presentation. (*Indian Dairyman*, Vol. 54, No. 2, 2002, p. 55)

DRIVERS OF GROWTH IN DAIRY INDUSTRY

In the increasing competitive dairy market, catering to traditional consumption by focusing on breakfast, lunch and dinner would not be sufficient to guarantee success and profits. Even as more manufacturers target the modern consumer, the reliance on new product development will increase because their product portfolio will experience unprecedented change.

Growth strategies in dairy must focus on the three trends driving market (A) health and nutrition needs, (B) pleasure and fun desire, (C) product presentation and packaging features.

Thus identifies fast growth opportunities each product group (52.80% recent august launches have made health claims), illustrated by the analysis of over 1200 innovative product launches from data monitor's world-wide Innovations Network. It also explores the impact of changing consumption habits on the industry.

DAIRY PRODUCTS AS THE CONSUMER'S FAVOURITE HEALTH FOOD

The demand for health products is stronger than ever before and predicted to be the most impelling factor in the dairy market within the next five or ten years. Since consumers are becoming increasingly aware of myriad health problems, the number of food products designed to alleviate or prevent them is increasing in an efforts to meet the emerging requirement. In addition to a number of historically well published health risks such as high cholesterol, obesity and heart disease many consumers face other problems like vitamin deficiencies and poor digestion. The issue of pesticide/veterinary drugs/ antibiotics resides will also leave its impression on the marketing of dairy products.

KEY MOTIVATIONS FOR DAIRY PURCHASES

As consumer depend increasingly on the food they eat to improve their health and well-being, manufacturers are responding with products which will, they hope, place them at the cutting edge of new product development.

Emerging from Japan are a variety of innovations containing ingredients designed to improve the condition of the skin and hair. New demands from consumers will also pose challenges for policy-makers and food law authorities. They need upgradation since science has launched all levels of the decision-making process. Unscientific justifications and decisions can be questioned under the WTO framework and cause embarrassment to the authorities in international forums. At the same time, any delay in domestic research, development effort and legal provisions for promoting innovation will be counter productive in the fast changing competitive global market.

CHALLENGES FOR POLICY-PLANNER

Growth will occur is milche segments and under exploited target audiences. As traditional 'health' products in other markets develop a stronger cosmetic focus. This particular innovation may prove popular, especially among body conscious consumers. For example, vitamin supplements are currently available in combinations blended specifically to target individual cosmetic needs such as clearer skin or stronger hair. The use of ingredient proves the way for foods proposing cosmetic benefits. Young women, for instance, rarely perceive that they need products to help the digestive process or lower Cholesterol.

The key to successful health innovations is found not only in adding value, but also correct positioning. Thus, Indian dairy industry must move away from traditional approach to innovative product formulations and business practices.

Notes and References

1. Setty, H.B.M. (1979) : *'Major Role for Co-operatives'*, Kurukshetra, March, 1979.
2. Singh, R.K.P. & A.K. Chaudhary (1999) : *'Dairy Development through Co-operatives'*, Spellbound Publications, Rohtak.
3. R.S. Jalal : *'Rural Co-operatives in India'*, Anmol Publication, 1996, p. 1.
4. Sharda, V. : *'The Theory of Co-operation'*, Himalaya Publishers, New Delhi, 1986, p. 1.
5. *Ibid.*
6. B.S. Mathur : *'Co-operative in India'*, Sahitya Bhawan, Agra, p. 2.
7. R. Rajgopalan (ed.) 1996 : *'Rediscovering Co-operation, Co-operatives in Emerging'*, Context, Vol. 3, IRMA, Anand, pp. 294-20.
8. Mohan, N. (1972) : *'Milk Co-operatives in India'*.
9. Shah, M.K. : *'Planning of Milk Supply for a Dairy Co-operative'*, Indian Dairyman, 19, 225, 1967.
10. Patrick, John : *'Economics of Dairy Development in India'*, Prabhat Prakashan, Patna, Chapter-3, Section-4, p. 267-74.
11. Batra, B.D. (1979) : *'Dairy Co-operative for Accelerating Rural Development'*, *Kurukshetra.*

    Baviskar, B.S. (1990) : *'Dairy Co-operatives & Rural Development in Gujarat'*, M. Doornobs & K.N. Nair (Ed.) Resource Institutions.
12. Rao, V.M. (2004) : *'Empowering Rural Women Through Dairy Co-operative'*, Anmol Publications, New Delhi.
13. Amrita Patel (2004) : *'Indian Dairyman'*, Vol. 56, No. 10,
14. *Dairy India*, 1992.
15. Shekhawat, Bhairo Singh (2004) : *'Indian Dairyman'*, Vol. 56, (10).
16. *Ibid.*
17. Chakravarty, R. & Chand, R. : *'Indian Dairyman'*, Vol. 56, No. 11, 2004, p. 51.
18. *'Indian Dairyman'*, 2004, *op. cit.*, p. 54.
19. *'The Hindustan Times'*, Sunday Magazine, 10 May, 1988, p. 9.
20. *'Indian Dairyman'*, Vol. 56, No. 10, 2004, p. 18.

21. Chakravarti, R. and Chand, R. : *'Connecting Technology and End-users in Dairy Sector'*, India Dairyman, Nov. 2004, Vol. 56, No. 11, pp. 53-54.
22. Prasad, J. : *'Principles and Practice of Dairy Farm Management'*, Kalyan Publishers, 1977.
23. Rao, C.K. (1966) : *'A Text Book of Dairy Science'*, S. Bhattacharya & Co., Calcutta.
24. Welstra, P., Gensto, J.J. (*et. al.*) (1999) : *'Dairy Technology'*, Marcel Dekker, Newyork, U.S.A., p. 127.
25. *'Dairy Development in India'*, An Appraisal of challenges and Achievements—Concept Publishing Company, New Delhi.
26. Sharma, V.P. & Sharma, P. : *'Trade Liberalization and Indian Dairy Industry'*, Oxford & IBH Publishing, New Delhi, p. 15.
27. Sharma, V.P. & Sharma, P. : *'Trade Liberalization and Indian Dairy Industry'*, Oxford & IBH Publishing, New Delhi, p. 163.
28. Sudhir Krishna : *'Indian Dairy Association'*, 2002, p. 43.
29. Nair, K.N. (1981) : *'White Revolution in India : Facts and Issues'*, Economic & Political Weekly, 20(25 & 26) A-95.
30. Mishra, S.N. (1979) : *'Livestock Planning in India'*, Vikash Publishing House, New Delhi.
31. Rao, V.M. (1983) : *'Towards White Revolution'*, Kisan World 10(8), 26-27.

6

Package Programme for Dairy Development and Operation Flood

CONCEPT OF OPERATION FLOOD

The IDC's (Indian Dairy Corporation) main activity was the implementation of the Operation Flood Project (OFP). The project was taken up by the Government of India in 1970. The basic concept of the project was the organisation of a well functioning of rural co-operative structure, which could effectively handle the provisions of inputs to improve milk production and the productivity of dairying. The work involved the indenting, receiving, movement, storage, quality control and issue to the milk plants the OFP gifted commodities to the tune of 1,26,000 tonnes of skim milk powder and 42,000 tonnes of butter oil in the Operation Flood-I project.[1]

India's operation flood or the *'white revolution'* as a project, is properly known is not only the world's largest dairy development programme but possibly the world's most controversial.[2] Indeed, the programme encompasses several points of agricultural economy and nutritional patter, the relevance of Euro-American dairy technology in tropical areas, the dangers of relying on dairy aid from the EEC (European Economic Community), and various conflicting interests at stake both within and outside the country.[3] (Shanti George, *EPW*, Vol. XX, No. 49, Dec. 7, 1987)

Shastri George rightly stated, "In so far as intra-national interests have come under scrutiny is the context of Operation Flood, the focus has generally been on the relationship between rural and urban India established through the dairy sector with producers' interests being identified with rural territory and consumer interests with urban areas.

The perspective on operation flood has stimulated debate on the flow of milk from villages into cities, the National Milk Grid through which the flow takes place, and the reverse traffic of money from urban to rural areas in exchange for milk."

We analysis the *'white revolution'* or 'Operation Flood into its components facets of production, procurement marketing and consumption in order to inquire about each of these components.

(a) Which rural class benefits from the production strategies of OFP that revolved around cross-breed cow fed on green fodder and compounded feed ?
(b) Which classes in rural areas find operation flood's procurement network of Anand pattern dairy co-operative advantages ?
(c) What is the impact on various village classes of marketing through operation flood's milk grid ?
(d) Which rural classes gain in terms of milk consumption from White Revolution ?

Operation Flood (OF) or White Revolution intends to launch nothing less than a flood of milk and the sources of this flood is expected to be India's villages, containing 75 percent of the country's population and 95% of the bovines. This statement would arose only reception at first reading for the low yielding cattle and meagre fodder resources of rural India suggest and unlikely source of any lactic inundation.

TRANSFORMATION OF OPERATION FLOOD

However, Operation Flood is ambitiously designed to transform milk production in India; some 10 to 25 million milch animals form the milksheds covered by the programme are to be upgraded through cross breeding to form a National Milch Herd and in order to provide fodder input necessary for enhanced dairy output in the Operation Flood milksheds feed milking plants will be set-up and two million hectares of irrigated area will be used for green fodder production. (Nair, 1981 : NDDB, 1977).

The study of Operation Flood shows that the Co-operatives have been quite successful in increasing production and procurement of milk in Gujarat as well as Bihar.

The milk producers including weaker sections, who have become the members of the milk co-operatives, can earn substantial side income from sale of milk in the villages. The income raised in this way due to Operation Flood has been quite useful in improving the economic and social status of the villagers in the dairy villages as compared to that in the control villages.

HYPOTHESIS

Hence, the hypothesis that the organisation of milk co-operatives can go a long way in enhancing milk production and improving economic and social conditions of the rural population, especially the weaker sections of Bihar, Gujarat and Haryana.[5] (*EPW*, XXI, 22, May 31, 1986, p. 965)

The milk percentage of the marketed surplus of milk provided by the landless is 14.68 and 50.74 in 'dairy' and 'control' villages. The dairy consumption of fluid milk is milliliters per capital among the landless in 215 in 'dairy' villages and 225 in 'control' villages.[6]

The milk producers who could not get advantages of milk societies had expressed their anger and outrightly denied the favourable impact of such milk co-operatives.[7] In the operation flood programme, the Indian Dairy Corporation (IDC) handled the commercial and financial matters and the National Dairy

Development Board (NDDB) looked after *technical matters*. The funds generated to the tune of about one billion rupees through the sale of donated commodities, viz. butter oil and skimmed milk power were disbursed to ten states, viz. Andhra Pradesh, Bihar, Gujarat, Haryana, Maharashtra, Punjab, Rajasthan, Tamil Nadu, U.P. and West Bengal and the Union Territory of Delhi. This has enabled in expanding the existing capacity of dairies in the four metropolitan cities of Bombay, Delhi, Calcutta and Madras to 1.5 million litres per day, establishment of 4 mother dairies with a total capacity of 1.4 million liters per day in Bombay, Delhi, Calcutta and Madras, 17 feeder/balancing dairies with a capacity of 3.4 million liters per day and chilling plants with a capacity of 400,000 liters per day. This has created the framework for a modern dairy in India and may be considered as successful use of food aid for development with wider international applicability.[8]

OPERATION FLOOD-I

The project succeeded in increasing milk supplies from Operation Flood-I areas to city dairies through village co-operatives to 2 to 3 times pre-project supplies. This constituted 74 percent of their throughput in December 1980, the rest being met by recombined donated skimmed milk powder and butter oil. During this period, the availability of whole some milk at relatively stable and reasonable price to city consumers improved substantially. The official data indicates that during 1970 to 1980, milk production in India increased significantly and there is evidence that much of this was due to the activities of the Operation Flood-I. This programme has given a great attention to stimulate the increased milk production by action designed to enhance the productivity of milch animals. Under its technical inputs programmes 7,730 villages, 262 veterinary mobile units and 3,052 artificial insemination stations are covered.[9]

The animal husbandry and dairying sector comes under the state for policy concerns. However, Central Government formulates the policies in this sector and implementation is largely left to the states. This sector has attracted the attention of the Government because it provides income and employment opportunities in the rural areas.[10]

The investment on animal husbandry and dairying programmes shows the significance given to this sector by the Government for increasing production and productivity as obvious from the investment pattern of animal husbandry and dairying during various Five Year Plan period (from First Five Year Plan to Eleventh Five Year Plans 2007-12). Although, this sector occupies an important position and the contribution to the national economy is significant, the plan investment made so far does not appear commensurate with its output and future potential of growth and development.

In the post-Independence period, modernization of dairy industry became a priority of the Government with the initiation of a planning process.[11] Due of the first policy initiatives just after Independence was the recommendation from the *Milk Sub-Committee* of the *Policy Committee on Agricultural* (1950) which resulted in a *City Milk Scheme* set-up in Delhi, which grew eventually to cover nearly 100 towns and cities by 1960 (Parthasarthy, 2001).[12]

These were essentially demand driven, consumer-oriented initiatives run by Government and could not complete with milk vendors, as there was no supporting strategy to cover milk producers in rural areas with remunerative prices and other required incentives in an integrated manner.

AREAS OF NATIONAL BREEDING POLICY, 1962

The State and Central Government formally put through a National Breeding Policy in 1962. This Breeding Policy covered :[13]

(a) selective breeding of the pure Indian cattle breeds,
(b) selective breeding of pure Indian draught breed of cattle for better drought animals,
(c) selective breeding of dual purpose breeds for improving both milk and work output, and
(d) grading up of non-descript Indian breeds with exotic breeds, cross-breed was also introduced in 1969 and Government implemented various schemes for

increasing milk production such as : (i) Key Village Scheme (KVS), and (ii) Intensive Cattle Development Project (ICDP).

However, during the two decades between 1951 to 1971, the milk production remained more or less stagnant and growth in milk production was nearly one per cent.

The strategy for organised dairy development in India was initiated in the late 1960s after establishment of National Dairy Development Board.[14]

Operation Flood is fully articulated programmed designed to build on the strong infrastructure laid by the earlier project. The second phase with on outlay of Rs. 4,855 million was launched in Oct. 1979. Under this programme, the European Community assisted the country by donating milk powder and butter oil worth Rs. 2,352 million and the World Bank lent a helping hand with a loan of Rs. 1,730 million on soft terms. The phase II of the Operation Flood covered 250 districts, grouped into 168 milksheds in 23 states and union territories with over 5 million farm families.

In 49,000 village milk producers' co-operatives, they are now in a position to sell on an average 8 million liters of milk every day after retaining 25% of it for their own consumption. It fetches them about Rs. 30 million per day, aggregating to Rs. 10,000 million per year. The programme has thus, geared to enable the country to establish a balance between milk supply and demand and in turn raising per capita consumption to 157 per day in 1986-87. The Operation Flood Phase III programmed seeks primarily to consolidate the extensive milk procurement and marketing base already built in the earlier phases.

MAIN OBJECTIVES OF OPERATION FLOOD

The Government of India launched the 'Technical Mission on Dairy Development' (TMDD) in August 1988 to support and supplement the efforts of Operation Flood Programme[15] and to enhance to rural employment opportunities with income generation through dairying. The stated objective of the mission and Operation Flood Programme were :

(i) To accelerate the pace of increasing rural employment and income through dairy development on co-operative lines.

(ii) To accelerate the pace of application and adoption of modern technology to improve overall dairy productivity.

(iii) To ensure greater availability of milk and dairy products.

(iv) To dovetail State Government Programmes in Animal Husbandry and Dairying, Poverty Alleviation, IRDP, etc. with that of the dairy co-operatives.

(v) To dovetail research programmes of Central Government Research Institutes, State Agricultural Universities (SAUs) and National Dairy Development Board (NDDB) for optimum results.

These Operation Flood Programmes I, II and III transformed the Indian Dairy Scenario drastically from an insignificant one to be world leader. But all this growth in dairy production took place largely under regulated market environment both domestically and externally.[16]

With the increasing globalisation and liberalisation of this sector the future of the Indian Dairy Sector is now at cross-roads. The liberalisation of Indian economy at the signing of Uruguay Round Agreement (URA) of GATT in 1994 and provision of WTO have opened up global competition in the dairy sector which have promoted Operation Flood Programmes.[17]

The Operation Flood programme was launched by National Dairy Development Board (NDDB) to develop a viable and self-sustaining dairy industry in 1970. The key objective of OFP was to create a strong network and linkages between procurement, processing and distribution of milk by the co-operative sector and thus, linking the milk producing villages with the major urban markets.[18]

The first phase of Operation Flood (OF-I) was launched in 1970. The second phase of Operation Flood (OF-II) was implemented between 1981 and 1985 and the third phase of

Operation Flood (OF-III) was launched on April 1985 to March 31, 1996.

AREAS OF OPERATION FLOOD-I

The OF-I was launched following an agreement with the world food programme. OF-I programme involved :[19]

(i) organising dairy co-operative at the village levels,
(ii) creating physical and institutional infrastructure for milk procurement,
(iii) processing of milk,
(iv) marketing and production enhancement programmes/ services at the union levels, and
(v) establishing dairies in India's major metropolitan cities.

During the phase OF-I many states created Dairy Development Corporations to build co-operative structures and develop the dairy sector, however, these soon acquired the bureaucratic character of public sector organisation.[20] (Shah, *et. al.*, 1995)

OF II[21] covered 22 States and Union Territories and OF-III was launched on April 1985 to consolidate the extensive milk procurement, processing and marketing infrastructure created under OF-I and OF-II and finally completed on March 31, 1996.

TABLE 6.1

Salient Features of Operation Flood Programme

Key Parameter	OFP-I	OFP-II	OFP-III
(1)	(2)	(3)	(4)
1. OFP launched on	01.07.70	01.04.81	01.04.87
2. Completion Date	31.03.81	31.01.85	30.04.96
3. Investment (Rs. Crore)	116.50	227.20	137.95
4. No. of Milksheds	39	136	170
5. No. of DCSs Set-up	13,270	34,523	72,744

(Contd.)

TABLE 6.1 (*Contd.*)

(1)	*(2)*	*(3)*	*(4)*
6. No. of Members (lakh)	17.5	36.3	93.3
7. Average Milk Procurement (m kg/per day)	2.56	5.78	11.0
8. Liquid Milk Marketing (lakh/per day)	27.8	50.0	100.0
9. Processing Capacity Rate (MTPD)	45.4	88.0	192.0
10. Metro Dairies (lakh/liters/per day)	29.0	35.0	72.8
11. Milk Dairying Capacity (MTPD)	340.0	507.0	990.0
12. Technical Inputs No. of AI Centres	4868	7802	10915
13. Technical No. of AI done/year	8,20,782	13,29,455	39,43,890
14. Cattle feed capacities (MT/per day)	1.65	3.29	4.80

Source : NDDB, 2000.

TABLE 6.2
Various States Covered under Operation Flood Programme

OFP-I	*OFP-II*	*OFP-III*
Andhra Pradesh, Bihar, Delhi, Gujarat, Haryana, Karnataka, M.P., Uttar Pradesh, Maharashtra, Punjab, Rajasthan and W.B.	Andhra Pradesh, Assam, Bihar, Goa, Gujarat, Haryana, Himachal Pradesh, J & K, Karnataka, Kerala, M.P., Orissa, Maharashtra, Punjab Rajasthan, Sikkim, Tamil Nadu, Tripura, U.P., W.B., Andaman and Nicobar, Pondicherry and Delhi	Andhra Pradesh, Assam, Bihar, Goa, Gujarat, Haryana, Himachal Pradesh, J & K, Karnataka, Kerala, M.P., Orissa, Maharashtra, Punjab Rajasthan, Sikkim, Tamil Nadu, Tripura, U.P., W.B., Andaman and Nicobar, Pondicherry and Delhi

Source : NDDB, 2000.

Dairy Development Programme in the organised sector in Bihar started during the Fourth Plan (1969-74). On the request of then Chief Minister of Bihar, National Dairy Development Board took over the Dairy Development activities in the undivided State of Bihar for the replication of Amul (Anand Milk Union Limited Pattern) in the Year 1978, under Operation Flood Programme Phase-I. Different districts of the Bihar were surveyed for the availability of surplus milks. During this survey it was found that the northern parts of Bihar was the surplus milk zone, whereas the southern part of undivided Bihar (now Jharkhand) was milk deficit zone. It was also noticed that the then South Bihar, comprising of big industrial cities like Jamshedpur, Ranchi and Dhanbad had tremendous potentialies for marketing of milk and milk-products.

Under Operation Flood-I, an arrangement for artificial insemination on the 'Amul patter' was to be encouraged to improve the quality of cattle.[22]

OPERATION FLOOD-II

The Governments of Bihar, Andhra Pradesh, Madhya Pradesh, Uttar Pradesh, Rajasthan, Haryana, Punjab, etc. signed the initial agreement with the Indian Dairy Corporation to implement the Operation Flood-II project in the State in 1980. On the request of Government of various states, NDDB deputed a team in January 1981 to conduct the impressionistic survey for generating basic parameters required for identification of the project area and preparation of a perspective plan for dairy development under OF-II.

Patna Dairy Project

Situated in a green farm area of 28 acres the Patna Dairy Project had been brought in Bihar in 1977 under the globally renamed effort called 'Operation Flood', the brain child of Dr. V. Kurien.

As a precursor to the implementation of OF-II in Bihar, the Patna Dairy Protect requested the IDC to appraise their project proposal for financing for a period of 3 years. The appraisal of the proposals was completed by the IDC in the month of April 1982.

Salient Features of Patna Dairy Project

The salient features of the project proposal for Patna Dairy Project were as follows :

(i) 425 DCS would be established/restructured in Patna, Nalanda, Bhojpur, Gaya, Saran and Vaishali districts.

(ii) Coverage of 71,000 animals under milk procurement and approximately 10% of these under AI programmes.

(iii) The project envisages to procure some 37,000 litres of milk daily by third year from its area of operation.

(iv) To process, the milk expected to be processed, it is proposed that the existing Patna dairy be strengthened and its expansion to 2 lakh litres per day be considered under OF-II programme.

(v) The existing cattlefeed plant would also be strengthened to operate at full capacity.

Patna joined the White Revolution Plan and earlier in its own peculiar brand.

The entire cost of the project, for a period of 6 years, was estimated at Rs. 681.40 lakhs, of which Rs. 58.96 lakhs will be financed under Patna Dairy Project for a period of 3 years and the balance would be considered under OF-II programme.

For implementation of OF-II programme in Bihar, it is proposed to group 22 districts identified into 9 milksheds.[23]

Salient Features of OF-II

The salient features of the OF-II consists of the following :

(a) Production of an incremental 5.86 lakh lpd of milk.

(b) Establishment of 4,000 'Anand Pattern' Dairy Co-operative Societies (DCS) including re-organisation of existing milk collection centres/conventional societies.

(c) Establishment/expansion of an incremental 6.40 lakh lpd processing capacity and 3.10 lakh lpd of chilling capacities.

(d) Construction/expansion of an incremental 200 MT capacity of balanced cattlefeed per day.
(e) Coverage of some 6.5 lakh milch animals under technical input programme and 1.66 lakh milch animals under artificial insemination.
(f) Contribution of some 2 lakh lpd of milk to NMG (one lakh lpd in tetrapak and another one lakh lpd by road/rail milk tankers.
(g) Training of approached teams, DCS and unions' staff..
(h) Provision of technical targets are given in Table 6.1.

Institutional Structure

The BSCMPF would be the implementing agency for the project. The federation would own and manage centralised facilities for the use of its member unions. These facilities would include :

(a) Frozen semen production station,
(b) liquid nitrogen plant and the delivery system,
(c) dairy plants and milk distribution facilities (including road milk tankers) in class-I cities of the State which are outside the project area viz. Ranchi, Jamshedpur, Bokaro and Dhanbad (including Bermo).
(d) cattlefeed compounding facilities, and
(e) diagnostic laboratory.

In addition to managing these facilities, the federation would be responsible for publishing extension materials, marketing milk outside the project area, marketing milk products, co-ordinating inter-union milk movement and supervising procurement of milk/milk products from NMG. The federation would also initiate formation of the proposed 9 unions, co-ordinate and guide them for their overall activities. IDC/NDDB would assist in recruiting and training the core staff of the BCMPF and its spearhead teams.[24]

FUNCTIONS AND RESPONSIBILITIES OF UNIONS

The main functions and responsibilities of the unions would be :

(i) organising DCS on Anand pattern,
(ii) training of DCS staff,
(iii) arrangement for transportation of milk procured through DCS, its processing and marketing in its area of operation, regular payment to the DCS for the milk procured,
(iv) providing technical inputs such as animal health and AI services, cattlefeed, etc. required for increased milk production,
(v) arranging continuous and concurrent audit of the accounts of DCS, and
(vi) extension activities like organisation of dairy and fodder demonstration farms, educating the producers in improved animal husbandry techniques by arranging field visits, etc.

MAIN FUNCTIONS OF DCS

The main functions are responsibility of the DCS would be :[25]

(a) Daily collection of milk from the member producers.
(b) Recording and testing of milk and regular payment on the basis of the composition of milk.
(c) Provision of micro-level technical inputs like AI services, first-aid, etc.
(d) Sale of cattlefeed to the milk producers.
(e) Provision for extension services regarding organisation and operation of co-operative services and technical information for improved animal husbandry practices.

Project Costs

The entire project (excluding PDP) was estimated to cost Rs. 56.28 crores. The internal rate of return on the investments under the project was worked out to 14.95%.

Thus, it is clearly evident that dairy development programme in the country has gone through a metamorphic change from the initial key village scheme and Intensive Cattle Development Project aimed at upgradation of local cattle in selected tracts to a nationwide anti-poverty programme as instrument to social security in the rural India and Bihar.

In the meantime in 1989 technology mission on Dairy Development was launched to argument the rural income through the application of modern technologies in dairying. It also emphasized the optimization of resource use for cattle improvement in terms of breeding farms, bull mother farms, semen freezing and artifical insemina facilities, veterinary health care facilities, etc. available with the Bihar State Animal Husbandry Department under Operation Flood.

In addition, the research establishments under the Indian Council of Agricultural Research (ICAR) and State Agricultural Universities (SAUs) have also been linked up in the process to provide the research/technological support.

Presently, the dairy development component has become a part of the countries rural development programmes with extension to schemed like Training of Rural Rank to Self Employment (TRYSEM), Samll Farmer Development Agency (SFDA), Tribal Welfare Programme, etc. to ensure food security to weaker sections through self-reliance.

The success of Operation Flood Programme brought up India's milk production from 22 million tonnes in 1970 to 88 million tonnes in 2002. Per capita milk availability also grew 114 gram/day in 1970 to 230 gram/day in 2002.

TABLE 6.3

Union-wise Details of Physical Target under OF-II in Bihar

Sl. No.	*Name of Union*	*DCS organ.*	*Animals under procure-ment ('000)*	*Animals under AI ('000)*	*Avg. milk procure-ment ('000)*	*Dairy plant ('000) LPD)*	*Chilling centre ('000) LPD)*
	(1)	*(2)*	*(3)*	*(4)*	*(5)*	*(6)*	*(7)*
1.	Patna	600	90	22	62	300 (5MT)	20
2.	Rohtas	600	105	25	73	—	89
3.	Gaya	400	60	16	43	60	40

(Contd.)

TABLE 6.3 (*Contd.*)

(1)			*(2)*		*(3)*	*(4)*
4. Saran	400	62	15	46	—	60
5. Muzaffarpur	300	51	14	25	60	40
6. Begusarai	600	114	31	67	100 (5MT)	40
7. Monghyr	400	59	16	48	60	—
8.. East Champaran	400	49	12	38	—	60
9. Saharsa	300	59	15	20	—	30
Total	4000	649	166	422	580	370
City Dairies						
1. Jamshedpur					125	
2. Bokaro					50	
3. Dhanbad					100	
4. Ranchi					60	
Total					335	
Total Dairy Capacity					915	

Figures in paranthesis indicate powder plant capacity.

DISPARITIES

But due to operation flood there are still large disparities in milk availability—

TABLE 6.4

Years	*1960*	*1970*	*1980*	*1990*	*1995*	*2000*	*2001*	*2005*	*2010*	*2020*
Milk production (Million tonnes) (Projected)	20	22.2	31.6	53.9	60.2	81	88	98	120	180
Per Capital availability	124	114	128	178	197	219	230	249	270	315

Source : Dairy India Year Book.

(a) The rural population which accounts for 72 percent of the country's population and most of the milk production, retains only 46 percent of the milk for the consumption. Rural per capita consumption of 144 gm/day is a third of the urban consumption of 437 gm/day.

(b) There are disparities across income groups. The lowest income group consumes 44 gm/day. This is one-tenth of the consumption of the highest income group 416 gm/day.

(c) Per capita milk availability across states varies from 14 gm/day to 945 g/day.

RELEVANCE OF OPERATION FLOOD

As we know, milk is an important part of the Indian diet. We have been consuming milk and milk products for thousand of years. Even our ancient scriptures give references of the health benefits of milk and milk products. Milk and milk products account for 17 percent of household food expenditure.

Operation Flood (OF) is a prerequisite to assessing the instruments that evaluate dairy development programme. Operation Flood is a path-breeding approach to dairy development in India, and indeed has inspired a desire for replication elsewhere in Asia and Africa.[26] (Shanti George, *EPW*, Vol. XXI, No. 23, June 7, 1986, p. 1020).

Operation Flood attempts a huge increase in milk production in India through the innovations of cross-breeding cows with exotic strains diets composed mainly of green fodder and factory compounded feed, supplemented by enhanced Veterinary protection.

The Anand pattern of dairy co-operative that originated in the Khaira, district of Gujarat replicated in the milksheds covered by operation flood. These co-operatives are to procure milk from member producers, and channel income and inputs.[27]

ISSUES OF EVALUATION OF PACKAGE PROGRAMME

Under the operation flood programme, rural milk is to travel to city homes through a chain a chilling centres, feder-balancing plants of which some will manufacture products, urban dairies and bulk vending units, with milk tankers linking these units by rail and roads.

Evaluators must therefore, assess the effectiveness of the infrastructure. In addition, they have to test that which funds the programme and that which result from it. A large part of the funding comes from the recombination and sale of dried skimmed milk and butter oil donated by the European Economic Community (EEC) from its dairy surpluses, and evaluators must consider whether their donations are being used in a manner that will ultimately lead to self-sufficiency in the Indian dairy sector or whether this sector in becoming so dependent on donations that the end-result of the strategy will be commercial imports from the ECC.[28]

Another vital issues for evaluation is which producers and consumers benefit from the programme and whether there are needy groups from the point of income in the case of producers and nutrition in the case of consumers.

Such are the major areas of Operation Flood Programme on which evaluators must deliver their deeply considered judgement, in addition to the performance of the implementing agencies which are the National Dairy Development Board (NDDB) and Indian Dairy Corporation (IDC).

Operation Flood has been regularly monitored and evaluated by the sponsoring agencies. Dairy commodities to fund the first phase were channelled to India via the World Food Programme (WFP) of the Food and Agricultural Organisation of the United Nations. UN inter-agency mission visited in 1972, 1976 and 1981, the third time to conduct a terminal evaluation of OFP-I since the WFP's association with the project ended with the first phase.

In addition to monitoring by these international sponsors, the Government of India appointed in February 1984 an Expert Committee headed by L.K. Jha to evaluate the performance of the IDC and NDDB in terms of the specific objectives of Operation Flood-II. The Jha committee handed its report in December 1984.

Similarly, Operation Flood Programmes were evaluated with regards to : (a) the procedures adopted, (b) the date generated, and (c) the conclusion arrived.

According to Shanti George (*EPW*, XXI, No. 23, June 7, 1988, p. 1021) it is unrealistic to expect international mission and national committees to monitor development under operation flood an all levels of region, state, district, taluka, metro city, city and villages.

Indeed, when we flip through the list in report of individuals with whom international missions and national committees meet, we find that they consists mainly of chairpersons, deputy directors, managing directors, general managers, commissioners, advisors, members of parliament.

An overall evaluation must be extensive, intensive, rigorous, balanced and the cumulative result of the similar evaluations at the various levels subsumed.

According to WFP Mission Report, "In India, the rural poor including the landless labourers very frequently keep dairy animals, one of three per household being common." (FAO 1981: 76). The FWP Mission took into account criticism of Operation Flood. Among the Mission's terms of reference is the following section : "to comment on the validity of various public criticisms of the OFP for example are:

(i) Operation Flood has made the country dependent on imported milk powder.
(ii) Operation Flood has made the country more dependent on imports of dairy machinery and the imports of new technology.
(iii) Operation Flood has benefitted the urban and rural rich.
(iv) Operation Flood is an 'expensive project'.
(v) Operation Flood has attempted to produce milk in India when all the land and its resources were really required for feeding the growing human population.
(vi) Operation Flood has not increased milk production in India, but has only built dairy factories which are languishing for milk.

Item (vi) refers to milk production enhancement and capacity utilisation of dairy plants. On the first of these subjects the Mission reports, "it is difficult to assess precisely the extent of production increases, it would be useful to monitor development of milch animals and production more systematically on the level of district unions and village co-operatives.

The unorthodox approach adopted by this project is noteworthy, dairy processing and marketing has been given higher priority over dairy production technology.

To after refer World Bank Report, 1978 : 31, "dairy import would continue to be necessary to supplement the production of milk powder and butter oil required for recombination in the summer months of minimal milk supplies. Buffer stocks of skimmed milk powder and butter oil also need to be created to allow."[29]

The fact that, notwithstanding this large increase in overall milk procurement in the operation flood area, the utilisation of dried skim milk and butter oil has been phased out more vigorously in recent years has to be seen is the context of the expansion of Operation Flood (FAO, 1981 : 58).

Moreover, development and growth of dairy sector in India has been fascinating. During 1960s, India imported 43% of its consumption of milk solids and now she is a net exporter. This has been achieved by 'Operation Flood Programme' acknowledged internationally and by the World Bank for its phenomenal success. The core feature is the co-operatives with farmer control at milk procurement, processing and marketing. Strength in marketing and brand development helped realise better price for milk paid to the farmer and incited him to invest in milk production. Apart from Amul known co-operative brands are Vijaya, Verka, Milma, Aavin, Nandini, Saras, Parag, Vita—to name a few. They have withstood the competition with multinational brands.[30]

The Operation Flood Programme provided a framework for farmers to reach the urban markets. This framework which is famous as the 'Anand pattern' essentially structured the farmers into 2 tier co-operative structure. By end of phase III of Operation Flood, 9 million farm families were linked to 70,000 primary co-operatives.

Operation Flood-II was financially supported by both EEC (dairy commodities aid) and the World Bank (financial oil). In 1986, the Operation Flood-III was proposed by the Government of India to ECC and World Bank to further extend and consolidate the dairy development in the country of strengthening the grassroots level infrastructure facilities for milk procurement, milk processing and marketing network established under Operation Flood-I and II.[31]

The Operation Flood-III was proposed upto 1994 which is mainly intended to cover the imbalances remaining after Operation Flood-I and II with a coverage to traditional 15,000 villages and 200 medium size urban centres. The Operation Flood programmes thus, have been implemented with the aids and grants received from World Food Programme (WFP) of the United Nations Food and Agricultural Organisation (FAO), the European Economic Community (EEC) and the World Bank.

The World Food Programme (WFP) has supported the Operation Flood (1970-81) with its dairy commodity aid and financial grant involving a total outlay of nearly Rs. 1200 million and Operation Flood-II and III have involved an outlay of Rs. 4855 million and Rs. 6806 million respectively. But of the total funds received about one-third of the total cost of these projects have been funded by World Bank and rest (about 50%) originated in the form of dairy community and from EEC.

The Operation Flood Programmes by and large differed from ICDP that they laid emphasis on milk supply and marketing schemes and viewed dairy development as an instrument for rural development and social change.[32] This is because of the realisation of the fact that growth of dairying requires apart from the increase in the milk production, the milk production centres need to be linked with marketing centres to ensure farmers and customers their far deal and to create a sustainable strong base for the dairy development in the country and Bihar.

The Operation Flood, part of the dairy development, has clearly emphasized the social aspects of developing through the importance given for the establishment and growth of Village Milk Producers' Co-operative Societies (VMPCSs) throughout the country followed by the initial success of Anand Milk Producers' Union is uplifting the farmers. The details of the growth of Anand Pattern Village Milk Producers' Co-operation Societies the OFP is given in Table 6.5.

TABLE 6.5
Growth of Village Milk Producers' Co-operative and Procurement under Operation Flood (1970 to 1990)

Year	*Member VMPCS*	*Annual Milk Producers ('000)*	*Procurement (Million Tonnes)*
(1)	*(2)*	*(3)*	*(4)*
OFP-I			
1970-71	1588	278	0.1898
1971-72	1811	327	0.2372
1972-73	2200	361	0.2774
1973-74	2598	394	0.2226
1974-75	2966	445	0.3175
1975-76	4533	562	0.4197
1976-77	7681	723	0.5657
1977-78	9306	943	0.6205
1978-79	10099	1213	0.7336
1979-80	11436	1475	0.8614
1980-81	13270	1747	0.9844
OFP-II			
1981-82	18422	2125	1.0147
1982-83	23496	2620	1.6133
1983-84	28614	3116	1.9016
1984-85	34523	3632	2.1097
OFP-III			
1985-86	42692	4484	2.8762
1986-87	49077	5097	2.8652
1987-88	54525	5666	2.8105
1988-89	58883	6250	2.9090
1989-90	60825	7003	3.5821
1990-91	61901	7619	3.9841
1991-92	62700	8014	4.4871

(Contd.)

TABLE 6.5 (*Contd.*)

(1)	*(2)*	*(3)*	*(4)*
1992-93	64209	8267	4.9651
1993-94	65867	8413	5.1212
1994-95	68653	8567	5.2023
1995-96	70642	8745	5.8014
1996-97	72543	8912	5.8141
1997-98	73775	9016	5.9203
1998-99	75448	9029	5.9886

Source : Compiled from Various Records Reports of COMPFED of Bihar Unions to reach more than 600 cities and towns.

Phase I : 1970-81

Phase II : 1981-85

Phase III : 1985-96

SCOPE OF OPERATION FLOOD

Under Operation Flood Programme, there exists some scope for improving the overall efficiency, usefulness and popularity of milk co-operatives for the benefit of certain groups of the people of the villages.

Thus, obviously it can be recommended that—

1. All the landless people and small farmers including Harijans and other lower caste people who own milch animals, should be convinced to become members of the milk societies.
2. The milk society also should be asked to accept such milk producers as members provided they fulfil the requirements for getting membership.Their share capital may either be paid for development fund or through advances to be recovered from the proceeds of milk later on.
3. The case of avoidance of membership to the low caste people should be spotted out by the supervisor of the milk union so that corrective steps can be taken.

BENEFITS OF OPERATION FLOOD

Moreover, in Operation Flood Programme the milk co-operatives are of considerable benefit to the weaker sections in the villages, provided these people own milch animals and their milk is accepted by milk co-operations. In practice, there are certain difficulties which seem to have inhibited these people to take due advantages of the milk co-operatives in the dairy villages.

No doubt, most of the landless households selling milk have taken up production and sale of milk due to the dairy co-operative movements, the largest beneficiary group is that of the Christians community only a very small proportion of landless households does not sèen to have as yet succeeded in the sample villages.

SUMMING UP

Thus, Shanti George has observed in his studies (*EPW*, Dec. 7, 1985, p. 2168), "Milk has been treated as a cash crop bringing income to the producers', households but it is also a food crop, a portion of which should be not sold but consumed." This is so for a number of reasons :

(i) In India sizeable amount of the proteins required as human diets is provided by pulses and lentils but more appropriately obtained from milk.
(ii) Milk is especially essential for the healthy development of children under five years of age and for the well-being of the women who are pregnant or lactating.
(iii) Milk is required in all rural homes.
(iv) Thus, operation flood is to be applanded when it takes of National Diet that includes 180 gms. of milk per person.
(v) The per capita availability of milk in India is at the most optimistic estimates only 135 gms. per day.

However, Operation Flood does not have an elaborate policy or the concrete infrastructure with regard to milk consumption, as the programme does in the case of milk production.

During pre-Operation Flood, the conversion of fluid milk into ghee within the producer's household yields a dairy product 'ghee', that is easy to cheap to store and transport. It also generate part of the proteins and minerals present in the milk as well as some butter fat. This by-product is highly perishable and also unlike ghee has little ritual or prestige value.

It is, therefore, generously given away to lower clashes, domestic servants and those without milch animals or whose animals are dry.

In fact, the prestige associated with butter milk is not that of consuming it in substantial quantities as with ghee but of giving it away plentifully and without charge, as the following description from a village in Bihar.

As stated in *Kurukshetra*, December 2005, pp. 30-33 Milk production in Bihar is dominated by small and marginal landholding farmers and by landless labourers who, in aggregate, own about 70% of the state's milch animal herd. These farmers in Bihar maintain an average herd of one or two milch animals, comprising cow and/or buffaloes. Dairying, as a subsidiary source of income, is a raw relief to most of these weaker groups of society. Dairying is an economic activity. It provides for an indirect income insurance against risks from crops such as crop failure due to drought or pests. The success of dairy development programme for Operation Flood in Bihar and other states has shown how ford aid can be used as an investment building the type of institutional infrastructure that can bring about COMPFED. The Programmes like Operation Flood with similar policy orientations, may prove to be appropriate to dairy development in India.

The big landliners who owned large of milch cattle, produced milk and ghee on large scale, give free of cost the butter milk to the landless and poor people. It is only due to the fact that butter milk in the villages has no market value and it is considered socially undesirable.

Operation Flood is modern and it does not procure traditional communities like ghee, leaving behind buttermilk. It instead, deals with fluid milk thereby taking every nutrient present in milk away from villages.

The new image of a big land owner in the White Revolution milksheds must be that of a man with several well fed cross-breed cows, whose female relatives carry many containers of milk to the village dairy co-operatives twice a day. It is now Operation Flood rather

Notes and References

1. Nair, K.N. (1979) : *Alternative to Operation Flood-II Strategy*, Economic and Political Weekly, 16(52), 21-29.
2. Bowonder, B.; Gupta, B.D.; Gupta, Sanjeev and Prasad, S.S.R. (1987) : *Further Evidence on the Impact of Dairy Development Programme*, Review of Agriculture, March, 28, 1987.
3. George, S. (1985) : *Operation Flood and Rural India, Vested and Divested Interests, EPW*, Vol. XX, No. 49, Dec. 7, 1985.
4. *Ibid.*
5. *Economic and Political Weekly*, 1986, Vol. XXI, No. 22, May, p. 965.
6. *Ibid.*
7. Singh, S. (1979) : *Operation Flood : Some Constraints and Implications*, Economic and Political Weekly, 14 (42-43), 1979, pp. 1765-74.
8. Sambhani, S. (1980) : *Transforming the Rural Poor : The Big Push in Action (The Experience of Operation Flood)*, Institution of Rural Management, Anand, 1980.
9. Mishra, M. (1987) : *The Land of the unfinished Revolution and the White Revolution*, 1987.
10. Shah, T. (1989) : *Economic Impact of Operation Flood*, Mimeo, Institute of Rural Management, Anand, 1989, p. 55.
11. George, S. (1985) : *Operation Flood*, Oxford University Press, 1985, p. 320.
12. Nair, K.N. and Keckson, M.G. (1981) : *Alternative to Operation Flood-II Strategy*, Economic and Political Weekly, Vol. 15, No. 52, 1981, pp. 2129-32.
13. Alvate, S.C. (ed.) (1985) : *Another Revolution Fails*, Ajanta Books International, Delhi, 1985, pp. XV + 274.
14. Elergo, R. (1981) : *Cost Analysis of Dairying*, Eastern Economist, 77(9), 422-23.
15. Jain, V.H. (1990) : *"Operation Flood : Constraints & Potentialities in Saurashtra"* in K.N. Nait, *Resources Institutions and Strategies : Operation Flood and Indian Dairying*, Sage Publications, New Delhi.
16. *Ibid.*
17. John, V.H. (1983) : *Operation Flood and Rural Poor*, Institute of Development Studies, U.K.
18. Baviskar, B.S. (1975 : *Operation Flood and Social Science Research*, Economic and Political Weekly, 1975.
19. Nair, K.N. (1980) : *Operation Flood : Some Constraints and Implications*, Economic and Political Weekly, 15 (8), 1980, pp. 446-48.

20. Shah, T.; Tripathy, A.K. and Desai, M. (1980) : *Impact of Increase a Dairy Productivity on Farmers' use of Feed Shiffs*, Economic and Political Weekly, Vol. 15, No. 33, 1980, pp. 1407-12.
21. Singh, S. (1981) : *Operation Flood-II : Some constraints and Implementation*, Economic and Political Weekly, 1981, Vol. 16, No. 32, pp. 2129-32.
22. Singh, P. Shekhawat (1987) : *Operation Flood in Rajasthan, Replication and Institutional Issues in a Regional Context*, 1987.
23. Banerjee, A. (1983) : *Operation Flood : Its Achievement in South India*, Tamil Nadu Journal of Co-operation, 74(889), 479-181.
24. Nair, K.N. (1981) : *An Alternative to the Operation Flood-II strategy working*, Paper No. 134, Centre for Development Studies, Trivendrum, 1981.
25. Nair, K.N. (1987) : *Operation Flood Re-examination : Report of Workshop*, Economic and Political Weekly, Vol. XXII, No. 7, Feb. 14, 1987.
26. George, S. (1985) : *Operation Flood*, Oxford University Press, 1985, p. 320.
27. Singh, Katar (1984) : *Impact of Operation Flood at the Village level*, IRMA, 1984.
28. Chaudhuri, A.K. (1969) : *Farming for Milk*, Indian Dairyman, Vol. 21, No. 73-83.
29. Mishra, M. (1985) : *Impact of Dairy Co-operative in selected areas of Gujarat*, 1985.
30. Verhagen, M. (1985) : *Operation Flood and the Rural Poor : Total Role of NGOs*, 1985.
31. Ramchandra, P.G. (1977) : *Some Imperatives for Bihar and Indian Forestry*, Final Report on the Bihar Forestry Agreement, OPTIMA, New Delhi, 1977.
32. Indian Dairy Corporation (1983) : *Operation Flood : A Reality*, Anand Press, Anand.

7

COMPFED and Second White Revolution

The word *'Bihar'* finds mention in the Vedas, Puranas, epics, etc. which has been coined from Buddhist monasteries *'Vihara'*. The Term *'Vihara'* was the chief scene of the activities of *Gautam Buddha,* and the 24 Jain *Tirthankars.*[1] According to S. Mishra "Earliest known habitation in the entire Ganga basin the oldest university *'Nalanda University'*, the world's most ancient highway the nuclei of the First Empire and second civilization of the Indian subcontinent, the earliest of the cave temples, the world's largest fair, etc. are found in Bihar."[2] Agricultural settlements in Bihar has been found from before 2000 B.C. by the Archaeologists. Cities emerged in the area around 500 B.C. During this period the ancient Indian State of Magadh dominated in region.

V.M. Rao[3] has rightly stated that "It was in Bihar that Buddha attained enlightenment, and that golden age of Mauryans existed. Great rulers of the state before *Christian era* were *Bimbisar, Udayain* (who founded the city of Patliputra)

Chandragupta Maurya and Emperor *Ashoka* of *Maurya dynasty*, the Sung and the Kanvas. Then came *Kushan* rulers followed by Chandragupta Vikramaditya of the Gupta dynasty. Muslim rulers made invoads into the territory during medieval period."

The first conqueror of Bihar was *Mohammad-Bin-Bekhtiyar Khilji*, Tughlaqs and Mughals followed who retained it until the British won battle of Buxar in 1764. At that time Bihar was part of Bengal but latter the two regions were separated.[4] Bihar became a province under British rules and declined into poverty due to policy of granting land ownership to local Jamindars. Bihar lost Purnia and Manbhum District to West Bengal during reorganisation of states in 1956 and the South Bihar (Chhotanagpur division) was renamed as Jharkhand in 2000. On November 14, 2000 Bihar was divided between Bihar and Jharkhand.

Jharkhand state comprises Singhbhum, Chhotanagpur, Palamau, Ranchi, Santhal Paragana, etc., i.e. the region of Jamshedpur (Tata); Deoghar, Dumka, Dhanbad, Bokaro, Giridih, Koderma, Hazaribagh, Chatara, Daltonganj, Loherdagga, Netarhat, Gumla, Simdega, Garhwa, Ranchi, Ramgarh, etc.

State of Bihar on other consists of Magadh Division, Patna, Sahabad, Bhagalpur, Munger, Muzaffarpur, Tirhut, Darbhanga, Saran, Champaran, Purnia Divisions, etc. Comprising the region of Gaya, Aurangabad, Jehanabad, Sasaram, Bhabua, Buxar, Arrah, Patna, Pali, Daudnagar, Nabinagar, Nalanda, Hilsa, Biharsharif, Rajgir, Hisua, Nawadah, Warshaliganj, Lakhisarai, Munger, Bhagalpur, Barh, Mukamah, Vaishali, Siwan, Chapara, Gopalganj, Motihari, Betia, Champaran, Hazipur, Samastipur, Rosera, Darbhanga, Laheriasarai, Madhepura, Saharasa, Madhubani, Supaul, Katihar, Purnia, Araria, Kisanganaaj, Nawgachia, etc.

Present Bihar after partition with Jharkhand is bounded in north by Nepal, in east by West Bengal, west by Uttar Pradesh and South by Jharkhand. The total geographical area of Bihar is currently 94,163 sq. km. constituting 2.86% of the total geographical area of the country and has 8.07% (8.29 crores) of total Indian population as per 2001 census.[5] Bihar has 37 districts comprising 533 blocks and 45,103 revenue villages. In this state about 7.1% of the total geographical area is covered under forests. Ganga, Kosi, Ghaghara, Gandak, Kamala Balan, Burhi Gandak, Sone, Phalgu, Durgawati, Karamnasa, Punpun are the main rivers of divided Bihar.

Bihar had a total geographical area of 173.88 thousand sq. millions, 39.43 millions are women with a density of population of 880 persons/sq. km. After bifurcation Bihar has 37 districts in which Gaya is the largest area while Patna has highest population. As of sex ratio, Siwan has 1033 females/thousand male population. On the other hand, Bhagalpur has the lowest sex ratio of (878).

Today Bihar is the poorest and most illiterate state in the country with the high levels of unemployment. Bihar is the most backward state of the country even in BIMARU (Bihar, Madhya Pradesh, Rajasthan, Uttar Pradesh and Orissa). Illiteracy and poverty make Bihar quite alarming for betterment of lives even if they have enough know-how. By weak and poor management problems are exacerbated and funds allocated for development, health, education and welfare are not properly utilized and accountability of impact is negligible.[6]

To quote Prof. Amartya Sen "and Jean Ire'ze the limited explanatory power of per capita income and related variable can be illustrated by considering the relationship between child mortality and the incidence of poverty in different states of India." The relevant information is being presented in the following figure which depict the state of affairs prevalent in the BIMARU State.

In Bihar there are several craft industries. In Mithila region (Madhubani, Darbhanga, Samastipur Districts) there are several craft industries being handled particularly by women. Traditional vegetable diets are obtained from leaves, mixed with goat milk, the point is then used on handmade paper. Women from the same district also weave a special grass called Sikki, they make brightly coloured boxes and baskets.[7] (World Bank, 1992)

BIHAR STATE DAIRY DEVELOPMENT CORPORATION

Bihar Dairy Development Corporation was registered under companies Act in the year 1972 for rapid implementation and controlling of dairy projects in Bihar. Like other states the management of dairy plants were shifted to the Bihar State Dairy Development Corporation (BSDDC) just like. This Corporation was set up in the Fourth Five Year Plan which emphasized consolidation and development projects of dairying in Bihar.[8]

The main objectives of setting up of the Bihar State Dairy Development Corporation were:[9]

(i) Increasing production of milk and milk products in Bihar.
(ii) Consolidation and expansion of existing dairy schemes to meet increasing demand for milk in cities.
(iii) Completion of spill over dairy schemes.
(iv) Establishment of milk depots in industrial towns.
(v) Establishment of rural dairies for distribution of milk in small towns and supply of surplus milk to the dairy factories.
(vi) Intensifying rural extension work including organization of milk collection and assembling centres for supply of milk to the dairy plants.
(vii) Increasing the drought capacity of bullocks by improved breeding and better feeding.
(viii) Providing adequate facilities for treatment of livestock and ensure protection by preventive measures.
(ix) Ensuring increased production of feeds and fodders.
(x) Training personnel in specialized fields.
(xi) Conducting research on livestock problems.

Thus, keeping in view the above objective BSDDC was made responsible for rapid development of dairying in Bihar and implementing Various programmes.

Table 7.1 demonstrates the livestock population trend between 1951 to 2001 in India Species-wise comprising cattle, buffalo, sheeps, goats, horses, camels, pigs, mules, donkey, Yarks, etc. Table 7.2 deals with per capita quantity and value of monthly consumption, rural and urban of liquid milk and milk products in Bihar. Table 7.3 shows district-wise estimates of adult female cattle and buffaloes in undivided Bihar containing of cross-breeds, local, adult female. Table 7.4 explains production and per capital availability of milk in India. Whereas

TABLE 7.1

Livestock Population 1951-2001 (All India Species-wise)

(*In million numbers*)

	1951	*1956*	*1961*	*1966*	*1972*	*1977*	*1982*	*1987*	*1992*	*1997*	*2000*	*2001*
(1)	*(2)*	*(3)*	*(4)*	*(5)*	*(6)*	*(7)*	*(8)*	*(9)*	*(10)*	*(11)*	*(12)*	*(13)*
Cattle	155.3	158.7	175.6	176.2	178.3	180.0	192.5	199.7	204.6	210.1	218.3	222.2
Buffalo	43.4	44.9	51.2	53.0	57.4	62.0	69.8	76.0	84.2	92.3	97.4	99.3
Total	198.7	203.6	226.8	229.2	235.7	242.0	262.4	275.8	288.9	302.4	316.1	321.5
Sheep	39.1	39.3	40.2	42.0	40.0	41.0	48.8	45.7	50.8	53.2	55.4	56.2
Goat	47.2	55.4	60.9	64.6	67.5	75.6	95.25	110.2	115.3	125.4	129.7	130.1
Horses and Pomies	1.5	1.5	1.3	1.1	0.9	0.9	0.9	0.8	0.82	0.86	0.89	0.89
Camels	0.6	0.8	0.9	1.0	1.1	1.1	1.08	1.0	1.03	1.05	1.06	1.09
Pigs	4.0	4.9	5.2	5.0	6.9	7.6	10.1	10.6	12.8	13.1	13.9	14.2
Mules	0.06	0.04	0.05	0.08	0.08	0.9	0.13	0.17	0.19	0.04	0.26	0.27
Donkey	1.30	1.1	1.1	1.1	1.0	1.0	1.02	0.96	0.97	0.98	0.99	0.99
Yanks	—	—	0.02	0.03	0.04	0.13	0.13	0.14	0.06	0.7	0.9	0.11
Total Livestocks	292.8	306.6	335.4	344.1	353.4	369.4	419.5	445.2	470.9			
Poultry	73.5	98.4	114.2	115.4	138.5	159.2	207.7	275.3	307.1			

Source : Livestock Census, Directorate of Economics and Statistics, Ministry of Agriculture, Government of India, 2002.

Table 7.2
Per Capita Quantity and Value of Monthly Consumption, Rural and Urban of Liquid Milk and Milk Products in Bihar

	Rural		Urban	
	Liquid Milk	Milk and Milk Product	Liquid Milk	Milk and Milk Product
Quantity (Litres)	2.41		3.40	
Value (Rs.)	23.62	25.77	42.61	47.70

Source : *Ibid.*

Table 7.3
District-wise Estimates of Adult Female Cattle and Buffalos, 1998

Sl. No.	District	Cross-breeds	Local	Buffalo	Total Adult Females
	(1)	(2)	(3)	(4)	(5)
1.	West Champaran	932	61341	64900	127173
2.	East Champaran	779	68917	120146	189842
3.	Sitamarhi	331	20217	63100	83648
4.	Madhubani	822	84102	168803	253727
5.	Kishanganj	136	76588	27600	104324
6.	Purnia	970	103717	76671	181358
7.	Katihar	270	85067	45076	130413
8.	Madhepura	270	65097	81880	147247
9.	Saharsa	20430	137543	143766	301739
10.	Darbhanga	1030	44936	76608	122574
11.	Muzaffarpur	1981	60985	1222444	185210
12.	Gopalganj	315	40834	73132	114281
13.	Siwan	310	43001	82098	125409
14.	Saran	786	55453	120956	177195
15.	Vaishali	1338	37978	98391	137707
16.	Samastipur	2191	72909	114593	189693

(*Contd.*)

TABLE 7.3 (Contd.)

(1)	(2)	(3)	(4)	(5)
17. Begusarai	1622	75078	67634	144334
18. Khagaria	169	41996	42928	85093
19. Bhagalpur	2553	90801	118994	212318
20. Munger	3545	177627	128407	309579
21. Patna	6820	61401	128407	309579
22. Bhojpur	5264	90593	222004	317861
23. Rohtas	1630	170899	255140	432169
24. Aurangabad	355	89017	96667	18609
25. Nawada	790	73215	59647	133652
26. Deogarh	2714	96409	19725	118848
27. Pakur	127	73041	15172	88340
28. Sahabganj	179	66324	29142	95645
29. Godda	189	108365	34009	142563
30. Dhumka Sadar	935	142514	18013	161462
31. Jamtara	275	73348	7023	80646
32. Ranchi	6160	93836	25512	125508
33. Khuti	477	98275	16216	114968
34. Lohardaga	616	26417	5149	32182
35. Simdegma	361	90625	8661	99647
36. Gumla	432	109697	23332	133461
37. Palamu	4372	132388	72428	209188
38. Garwah	194	103876	43027	147097
39. Koderma	340	69914	21161	91415
40. Singhbhum	1957	304569	22631	329157
41. Hazaribag	3240	224656	57117	285193
42. Dhanbad	3730	117598	28724	150052
43. Giridih	2464	204352	75869	282635
44. Bagha	709	61215	42131	104055

Source : Department of Fisheries, Animal Husbandry and Dairy Development (1998).

TABLE 7.4

Production and per Capita Availability of Milk in India

Year	*Production (Million Tonnes)*	*Per Capital Availability (gm. per day)*
1968-69	21.2	112
1980-81	31.6	128
1985-86	44.0	160
1990-91	53.9	176
1995-96	66.2	197
2000-01	80.9	222
2002-03	85.7	229
2002-03 (P)	89.4	234
2003-04 (A)	92.2	237

Source : Department of Fisheries, Animal Husbandry and Dairy Development (1998).

TABLE 7.5

Import and Export of Milk Products

(*in Metric Tonnes*)

Year	*Imports*			*Exports*		
	SMP	*B. Oil*	*Butter*	*SMP*	*Butter*	*Ghee*
1994-95	215	4304	—	5948	19	—
1995-96	1103	3858	—	2794	28	—
1996-97	283	—	37	447	142	—
1997-98	670	3460	—	1353	72	—
1998-99	1424	2608	51	655	180	538
1999-00	17252	5224	1381	2198	186	1015
2000-01	547	6382	18	3627	266	1318
2001-02	48	3243	1	14429	239	1425
2002-03	17	7478	4	11987	300	1544
2003-04	9112	3653	405	3070	197	906

Source : *Indian Dairyman*, Vol. 50, No. 10, 2004, p. 52.

TABLE 7.6
Registration under M and MPO (as on 31st March, 2004)

Sector	*Units (Numbers)*	*Capacity (Lakh litres/day)*
Private	395	356
Co-operative	202	316
Government	56	115
Total	653	787

Source : *Ibid.*, p. 53.

TABLE 7.7
Marketing of Liquid Milk—A Macro Estimate

1.	Estimated number of liquid milk brands	300
2.	Estimated sales of branded liquid milk	250 lakh liters/day
3.	Estimated rural milk marketable surplus	1260 lakh kg/day

Source : *Ibid.*, p. 53.

Table 7.5 deals with imports and exports of milk-production between 1994-95 to 2003-04. Table 7.6 demonstrates registration under M*MPO in hivate, co-operative and Government sector, Table 7.7 contains a macro estimate of marketing of liquid Milk. Table 7.8 states goes and projection to 2011-12, Table 7.9 highlights imports and exports of milk products in SAARC nations. Table 7.10 consists of per capita annual milk production and consumption in SAARC Nations whereas Table 7.11 deals with per capita milk production per annum and Table 7.12 indicates milk requirement and production in SAARC Nation.

A CASE STUDY OF BIHAR DAIRY DEVELOPMENT PROGRAMME

The Bihar State Dairy Development Corporation was set

TABLE 7.8

Goals to 2006-07 and Projections to 2011-12

1. Anticipated milk production in 2003-04	92.2 MT
2. Envisaged growth rate 2002-03 to 2006-07 6%	
3. Tenth Plan Targets for milk production in 2006-07	108 MT
4. Milk Production in 2011-12 6% CAGR on 04-05	147 MT
5. Milk Production in 2011-12 4% CAGR on 2004-04	126 MT

Source : *Ibid.*, p. 53.

TABLE 7.9

SAARC Nations : Milk Products Imports and Exports During 2001

(in 000 Mts.)

Country	*Milk Imports*	*Milk Exports*
Bangladesh	25.49	—
Bhutan	5.11	—
India	6.20	13.69
Maldives	1.68	—
Nepal	2.42	0.16
Pakistan	8.14	0.81
Sri Lanka	45.76	0.91

Source : Bhasin, 2003 and *Indian Dairyman*, Vol. 56, No. 10 (2004), p. 162.

TABLE 7.10

SAARC Nation : Per Capita Annual Milk Production and Consumption (kg.)

	1991		*2000*	
	Production	*Consumption*	*Production*	*Consumption*
Bangladesh	14.5	16.1	15.7	16.0
India	63.5	63.1	82.4	81.2
Nepal	50.8	48.2	52.0	50.0
Pakistan	134.1	117.4	191.1	154.8
Sri Lanka	14.8	34.9	15.7	41.9

Source : *Ibid.*, p. 103.

TABLE 7.11
Per Capita Milk Production

(*kg. per annum*)

States of India		*SAARC Countries*	
Haryana	236	Pakistan	191
Punjab	345		
Rajasthan	112		
U.P.	95	Nepal	52
Tamil Nadu	74	Sri Lanka	16
West Bengal	57	Bangladesh	16
Orissa	23		

Source : *Indian Dairyman*, Vol. 56, No. 10, 2004, *Ibid.*, p. 163.

TABLE 7.12
Milk Requirement and Production in SAARC Nations

Countries	*Milk Requirement (000 Mts.)*	*Milk Product (000 Mts.)*	*Surplus/ Deficit*
Bangladesh	10754	2112	–8642
Bhutan	156	N.A.	N.A.
India	85125	81790	–3335
Maldives	22	N.A.	N.A.
Nepal	1553	1171	–382
Pakistan	11543	26284	+14741
Sri Lanka	1595	292	–1300

Source : *Indian Dairyman*, Vol. 56, No. 10, 2004, *Ibid.*, p. 163.

up on 13th March, 1972 under the 1956 Companies Act.[10] The Bihar State Dairy Development Corporation came into existence with the objective of giving commercial orientation to the dairy industry in Bihar and improving the functioning of the dairy plants. A project of nearly Rs. 480 lakh was operated by the Corporation under the scheme of Operation Flood-I. Under this project, funds were mostly acquired from the Indian Dairy

Corporation, Baroda.[11] The Dairy Corporation was to develop the Dairy Corporation both at the village level and milk-shed level on 'Anand Pattern' and it was expected that the milk-shed level co-operative would take over the entire infrastructure created in due course.

The Indian Dairy Corporation (IDC) was incorporated as a wholly government owned company in Feb. 1970.[12] The immediate need for the setting up of the IDC was to handle the commercial transactions of the 'India WFP project 618' popularly known as Operation Flood. The IDC main objectives as per its Memorandum and Articles of association are to :

(i) promote the dairy industry in India,
(ii) assist the state governments and other organisations including co-operatives societies and private bodies interested in promoting dairy industry to meet the requirements of milk and milk products particularly to the vulnerable groups,
(iii) promote and develop cattle husbandry for increased milk production,
(iv) promote the removal from urban areas of high yielding milk cattle and calves, preventing thereby their premature destruction and their resettlement in rural areas,
(v) assist in developing and expanding the capacity and operations of existing dairies in the towns of Delhi, Bombay, Calcutta and Madras and other urban and rural areas,
(vi) assist in establishing and expanding the processing and handling capacity of liquid milk plants, chilling centres, feeding dairies, balancing dairies and all other facilities for increasing the processing and handling capacity of milk in the aforesaid four towns and in other urban and rural areas,
(vii) assist in setting up plants for producing and processing milk products,
(viii) assist in the manufacture and marketing of skimmed milk powder, butter oil and other milk products and by products thereof,

(ix) assist in erecting or providing storage and transport facilities for milk and milk products,

(x) act as agent of the Central Government for negotiating with the World Food Programme and other international and foreign agencies for obtaining skimmed milk powder, butter oil and other milk products, and

(xi) do all things necessary or advisable for carrying out the above objects.

The Bihar State Dairy Corporation after recruitment and training of necessary staff positioned the procurement and input wing (P and I) from 1975. A Spear Head Team (SHT) was deputed from National Dairy Development Board (NDDB) from the same year for helping the corporation in organising and developing the co-operatives. Though the progress in the initial years was encouraging, the corporation could not achieve the goals for which it was established. It has to incur losses in subsequent years in the following manners as indicated in Table 7.13.

TABLE 7.13
Profit/Loss of B.S.D.C.

Year (July to June)	*Profit/Loss (in Lakhs)*	
1972-73	–0.05	
1973-74	–1.75	
1974-75	–21.26	
1975-76	–20.46	
1976-77	–33.18	
1977-78	–94.60	
1978-79	–86.27	
1980-81	–93.00	
1981-82	–73.11	
1982-83	–32.08	
1983-84	–12.00	(Estimated)
Total	–591.44	

Source : Bihar State Development Corporation Limited, Patna, Annual Report, 1984-85.

Due to paucity of working capital and heavy establishment expenditure and operational losses the state government felt it necessary to request the Dairy Development Board for taking over the infrastructure created on management basis. The National Dairy Development Board (NDDB) took over the management of the infrastructure with effect from 1st Oct. 1981 under the banner of Patna Dairy Project (PDP).[13]

PROGRESS OF PATNA DAIRY PROJECT

Patna was identified under Operation Flood I as one of the milk sheds to feed the metropolitan city of Calcutta.[7] Accordingly a Feeder and Balance Dairy (FBD) and Cattle Feed Factory (CFF) were set-up in 1978 and 1977 respectively at Patna with a total investment of Rs. 297.8 lakhs.

The National Dairy Development Board (NDDB) immediately after taking over the project positioned an integrated Spear Head Team to restructure the milk procurement activities and also for streamline the working of the FBD and C.F.F. under the management of NDDB the project had not only made excellent progress but had been able to establish the fact the co-operatives could function equally well in Bihar too and what is essential is the proper atmosphere and guidance.[14]

Alongwith the organisation of milk procurement activities and management of both the plants on the commercial lines, the NDDB took special care to develop the Vaishal Patliputra Dugdh Utpadak Sahakari Sangh Ltd. (VPDUSS), the milk-shed level co-operative for taking-over the project once the Dairy Board withdrawn its management, NDDB handed over the Patna Dairy Project to Vaishal Patliputra Dugdh Utpadak Sahakari Sangh Ltd. (VPDUSS) with effect from Ist July, 1988.[15]

In order to establish dairy Industry on Amul pattern by involving farmers in organising the milk production and procurement, processing and marketing of milk produced in rural areas, the *Bihar State Co-operative Milk Producer's Federation Ltd. (COMFED)* was registered under the Bihar Co-operatives Societies Act, 1935.

Thus, for the implementation of dairy development schemes quickly in Bihar, the Bihar State Dairy Development

Corporation was set-up on March 13, 1972 under the 1956 Companies Act. The BSDDC was a Private Ltd. Company owned by the Government of Bihar. Before 1972, the Barauni Dairy unit was producing mainly ghee and butter but after BSDDC, it started to produce milk powder and ghee also. Under the Operation Flood-II a project of nearly Rs. 480 lakh was operated by the BSDDC. Under this project of Operation Flood-I funds were mostly acquired from Indian Dairy Corporation, Baroda. Out of its fund 70% of funds were in the from the loans and 30% were in the form of grants. Before COMFED, BSDDC had acquired Rs. 208.447 lakh in the form of loans and Rs. 83.333 lakh in the form of grants from Indian Dairy Corporation. With the help of these funds BSDDC had established a feeder balancing dairy at the cost of Rs. 6 lakh and cattle feed factory at the cost of Rs. 65 lakh. Nearly Rs. 30 lakh had been spent on technical inputs and establishing cattle breeding centre of foreign breed.

Under the same project there were 9 action items in which 5 action items were being implemented in Bihar. Under the action No. 4 and 5 a Feeder Balancing, Dairy of 1,00,000 litres procurement capacity was to be established for which there was a provision of Rs. 234.0 lakh which was envisaged to be spent mainly on the breed improvement of cattles, artificial insemination, cattle fed and other dairy developments. Under the action No. 8 there was a provision of Rs. 40 lakh for the establishment of Bull's farms. Under the action no. 9 there was a project to collect milk from rural areas, for which Rs. 26 lakh had been sanctioned.

However, the Bihar State Dairy Development Corporation BSDDC had manifold problems. As the milk procurement capacity of the corporation was inadequate, it was running in heavy losses. The sorry state of affairs through which BSDDC was passing could be curbed only when dairy milk procurement was sufficient to meet the existing capacity of dairies.

FACTORS INFLUENCED COMPFED

The inadequacy of milk procurement in Bihar during those

days particularly before formation of COMFED, was mainly attributed to the following factors:[16]

(a) There was a wide gap in the quantity of milk procurement between the leam season and flush reason. In the lean period there had a sharp fall in the procurement of milk.
(b) Another factor responsible for the low procurement capacity of BSDDC was the poor development of roadways in the rural areas of Bihar State. As most of the rural roadways were not communicated, any cost of communication to procure milk disappeared in the dairy season.
(c) There was a prevalence of middle men in rural areas, who did exploit the milk produces. On the one hand, they did purchase the milk at a less and sell that on a higher price.
(d) During these days, milk products were exported to other states and other places in bulk quantity dairy.
(e) Milk cows were sent to other states in a large number.
(f) There was a lack of technical personnel working in the dairy plants in Bihar.
(g) There was a wide gap between supply of the demand for milk in North and South Bihar (Bihar and Jharkhand).

As such, to overcome the financial and other problems of Corporations (BSDDC), the following steps were suggested and expected to be taken :

(i) Expansion of cemented roads in the rural areas was the pressing need to remove the procurement bottleneck of milk.
(ii) Middlemen had to be abolished.
(iii) To eliminate the middlemen, the process of the formation of milkman's co-operative societies had to be accelerated.
(iv) The export of milk products had to be banned. In

Rajasthan and Gujarat an arrangement of this kind had been made.

(v) The export of milk cows to other states had also to be banned.

(vi) Adulteration of milk by the middleness was expected to be discouraged.

Since, at the period of BSDDC milk production in Bihar was only 50 lakh litres per day which was only 80 grams per capita per day whereas our requirement was about 220 gm. per day. In those delicate situations, Bihar had not only to increase milk production but an effort had to be made to sell the milk at an appropriate price in the state.

In this context dairy development in Bihar and Bihar State Dairy Development had the following objectives in view[17] :

(a) *Milk selling at a Profitable price* : Milk selling in Bihar at a profitable price had to be managed for the milk producer living is far flung villages.

(b) *Supply of Milk at reasonable price* : In urban areas, milk and milk products had to be supplied at reasonable prices.

(c) *Supply of Pure Milk safeguard of Interest* : As consumers do not get pure milk. The production and the milk distribution have to be brought under organised sector, so that the interest of both the consumers and the procurements might be protected.

(d) *Supply of improved variety* : Milk producers had to be supplied on improved variety of feeder's seeds and feeder at a reasonable process.

(e) *Early loans to needy producer of Milk* : Loans had to be granted to the needy milk producers through nationalised banks to purchase milk cattle.

(f) *Training to villagers to procure fresh and unadulterated Milk* : Villagers had to be imparted a training to procure fresh milk free from any sort of adulteration.

(g) *Produce Cattle Insurance Scheme* : To provide first aid facility to cattle in rural areas farmers had to be benefited through introducing cattle insurance scheme.

Thus, keeping in view the existing position prevailing in Bihar. Bihar State Dairy Development Corporation was abolished and replaced by Bihar State Co-operative Milk Producer's Federation Ltd. (COMFED) in April 1983 to give a new dimension to the dairy sector in Bihar. It was only due to fact that the Co-operative Societies for selling the milk had been set-up on the pattern of Amul (Anand in Gujarat) whose membership was opened only for milk producers.[18]

But the result of COMFED in Bihar depends mainly on economic policy, rather than the existence of democratic and non-democratic form of government in the state. It is hightime that the intelligentsia of Bihar in particular and India in general pay some basic attention to improve the dairy farming in Bihar and the functioning of COMFED to meet the various challenges.

BSDDC laid emphasis on increasing milk production and development of drought animals. According to 1971 census, Bihar had 149.11 lakh cattle and 36.78 lakh of buffaloes. BSDDC had also laid emphasis in Bihar :

(a) to popularise practice of dairying among rural masses and to reach economic benefits to them by providing remunerative prices and various other basic technical inputs;

(b) to build up adequate and strong infrastructure in the potential milk-shed area, urban areas and at the apex level of the management so that, producers and farmers could be benefited within the stipulated period, for increasing milk production and its marketing through organised dairies;

(c) to organise adequate number of healthy milk producers' co-operative societies in the milk sheds and to provide economic relief to the producers through these co-operatives;

(d) to raise the installed capacities of the existing dairy plants and establish new dairy plants in the milk shed areas; and

(e) to set-up adequate number of milk marketing and distribution centres in the industrial towns.

Operation Flood Programme (OFP) was launched in Bihar

in 1977 and BSDDC was made responsible for implementing the programme of Operation Flood which aimed at improving milk marketing and raising procurement of milk and production in rural areas of Bihar. Initially it was taken up in Patna, Sahabad, Gaya and Saran districts. During Fourth and Fifth Five Year Plan various efforts were made to strengthen and consolidate BSDDC. By the end of 1979-80, milk production in Bihar increased to 18.6 lakh tonnes from 17.27 lakh tonnes in 1973-74. In this state during those days total 992 maternity hospitals and dispensaries were in functioning stage and per capita availability of milk in Bihar stood at 80 gram per day in the beginning of the Sixth Five Year Plan. During this Plan Fix Fluid Milk Plants at Patna, Ranchi, Bhagalpur, Barauni, Darbhanga and Bokaro, three Rural Dairy Centres at Khagaria, Samastipur and Fatuah 796 dairy Co-operatives and 6 Milk Unions were functioning.

BSDDC framed a plan during Sixth Plan which advocated integrated approach to dairy development in Bihar including production, collection, processing and marketing of milk and two tier dairy co-operative structure. BSDDC also laid emphasis on supply of technical inputs and services in the milk sheds. A Dairy Science College was also proposed to cater to the training requirements of the state.

The main aim of BSDDC was to work dairy farming in the pattern of Anand (Gujarat) Co-operatives in Bihar. But the performance of BSDDC was quite unsatisfactory.

Replacement of BSDDC as COMPFED

Thus, in order to replicate the Anand pattern dairy co-operatives, Bihar State Co-operative Milk Producer's Federation Limited (COMPFED) was constituted in April 1983 in place of Bihar State Diary Development Corporation by the Government of Bihar. All the plants were handed over the COMPFED in June 1984 and BSDDC was liquidated.

Mission of COMPFED

The role of COMPFED has been to organise farmers for forming dairy co-operatives and to make available remunerative market for their milk round the year. The mission of COMPFED is to transform rural economy through dairying in which the farmers are made instrumental for their own development and organised as a part of operation flood programme-II (OFP-II) with the Secretary, Department of Animal Husbandry and Fisheries of Government of Bihar as chairman.

TASK OF COMPFED

The tasks of the COMPFED are expected to:

(a) make policy decisions of milk products;
(b) to co-ordinate activities of milk unknowns in procurement, processing and marketing of milk;
(c) to guide milk unions in providing services like artificial insemination, veterinary first aid, Vaccination against diseases, feeding and management in operation flood;
(d) to establish 5000 Anand pattern dairy co-operatives;
(e) to increase of milk production at the rate of 5.86 lakh liters per day;
(f) to establish and expand an incremental 6.40 lakh liters/day processing capacity and 3.10 lakh of chilling capacity;
(g) to construct/expansion of an incremental 200 tonnes of balanced cattle feed/day;
(h) to establish milk marketing in the major towns at Patna, Muzaffarpur, Jamshedpur, Ranchi, Bokaro, Dhanbad, Gaya, Munger, Bhagalpur, Darbhanga, Biharsharif, Jamalpur, Arrah, Danapur, Chapara, Purnea, Katihar and Bermo for which an outlay of Rs. 15 crore was made available for the purpose during Eighth Five Year Plan.

ORGANISATION OF COMPFED

As per by-laws COMPFED has been constituted as:

1. M.D. Managing Director, COMFED — Chairman

2. Representative of the Ministry of HRD, Govt of India — Member
3. Representative of Diary Development, Govt. of Bihar — Member
4. Registrar, Co-operative Societies, Bihar — Member
5. Secretary, Welfare Deptt., Govt. of Bihar — Member
6. Managing Director, Women Development Corporation, Govt. of Bihar — Member
7. Project Director, Bihar Women Dairy Project — Member

COMFED (Bihar State Co-operative Milk Producer's Federation Limited) is functioning at the venue of Dairy Development Complex, P.O. Bihar Veterinary College, Patna.

Main Executive of COMPFED are :

1. MD (Managing Director).
2. Chief Manager-*cum*-Project Director.
3. MIS officers.
4. Associate Dean-*cum*-Director, Sanjay Gandhi Institute of Dairy Technology, Jagdeo Path, Patna.
5. MD—Desh Ratna Dr. Rajendra Prasad Dugdh Utpadak Sahkari Sangh Limited, Barauni Dairy, Barauni.
6. Incharge (P and I), Barauni Dairy, Barauni.
7. Incharge (Mahila Dairy Co-operative), Barauni Dairy, Barauni.
8. Manager, Barauni Dairy, Barauni.
9. Plant Manager, Barauni Dairy, Barauni.
10. Board of Director, Barauni Dairy, Barauni.
11. Supervisor, Barauni Dairy, Barauni.
12. MD, Dy. Manager, Asstt. Manager, Milk Procurement Officer, IEA, P and AH Assistant—Mahila Dugdh Utpadak Sahakari Sangh Limited, Samastipur Dairy, Industrial Area, Samastipur.
13. Manager, Procurement Officer, Ranchi Dairy Liquid Milk Plant, H.Ec. Sector Dhurwa, Ranchi.

14. Do—Sahabad Dugdh Utpadak Sahkari Sangh Ltd. Arval.
15. Tirhut Dugdh Utpadak Sahkari Sangh, Muzaffarpur.
16. Vaishali Patliputra Dugdh Utpadak Sangh, Patna Dairy Project, Fulwarisharif, Patna.
17. Bokaro Dairy, Bokaro Steel City, Bokaro.

OBJECTIVE OF COMPFED

COMPFED has been constituted replacing BSDDC in Bihar with the following objectives :

(a) for carrying out activities for promoting production procurement, processing, marketing of milk and milk products;
(b) to develop and expand other allied activities as may be conducive to promotion of dairy industry;
(c) to purchase commodities from members of other sources (without effecting interests of its members), pool, process, manufacture, distribute and sell the same;
(d) to study problems of mutual interests related to production, procurement and marketing of dairy products;
(e) to purchase and/or erect buildings, plant, machinery and other ancillary equipments;
(f) to establish research and quality control laboratories;
(g) to provide veterinary aid, artificial insemination, services and other animals' husbandry activities;
(h) to arrange training to staff members, unions and societies;
(i) to promote organisation of primary societies, and assist members in managing them; and
(j) to encourage fodder production by members of affiliated societies.

Bihar State COMPFED was 11 Board members in which 5 are nominated by the Government of Bihar :

Chairman/MD

Secretary, (An I.A.S. Officer), Department of Animal Husbandry and Fisheries, Govt. of Bihar.

Members

1. Secretary, Co-operative Societies, Bihar (IAS Officer).
2. Registrar, Co-operative Societies, Bihar (IAS Officer).
3. Representative for National Dairy Development Board, (A Technical expert from the field of Dairying).
4. Director, Diary Development (An elected farmer representative from Union).
5. Chairman, Vaishali, Patliputra Milk Union, Patna (An elected farmer representative from the Union).
6. Chairman, Barauni Milk Union, Barauni (An elected farmer representative from the Union).
7. Chairman, Mithila Milk Union, Samastipur (An elected farmer representative from the Union).
8. Chairman, Tirhut Milk Union, Muzaffarpur, (An elected farmer representative from the Union).
9. Chairman, Sahabad Milk Union, Arrah, (An elected farmer representative from the Union).
10. Managing Director, COMPFED (An Officer of Indian Administration Service, IAS).

FUNCTIONING OF COMPFED

Bihar State COMPFED works on a four-tier system. It was at the grassroot level, over 4000 Dairy Co-operative Societies in about 5000 villages covering 12.6% of inhabited villages of the State spread over 27 districts.

These Dairy Co-operative Societies (DCSs) are affiliated to five district milk producers' co-operative unions which in turn federated from COMPFED and a member of the National Co-operative Dairy Federation in India.

In Jharkhand districts of East and West Singhbhum, Ranchi, Bokaro and Dhanbad are being covered by dairies under control of COMFED for supply of milk and milk products to urban consumers. These dairies perform only marketing of milk and milk products supplied to them mostly from the North Bihar Milk Unions.

Milk Unions cover various districts of Bihar which can be shown for following Table 7.14.

TABLE 7.14
Coverage of District by Milk Unions COMPFED

Sl. No.	Milkshed Regions	Head Quarter	Coverage of Districts
1.	Mithila	Samastipur	Darbhanga, Samastipur, Madhubani and Katihar.
2.	Tirhut	Muzaffarpur	East and West Champaran, Sheohar and Sitamarhi.
3.	Sahabad	Arrah	Bhojpur, Buxar, Rohtas and Kaimur.
4.	Barauni	Barauni	Khagaria, Begusarai, Saharsa, Lakhisarai, Munger, Shaikhpura.
5.	Vaishali Patliputra	Patna	Patna, Nalanda, Siwan, Saran, Vaishali.
6.	Bhagalpur		
7.	Jamalpur		
8.	Ranchi	Under district control of COMPFED.	
9.	Gaya		
10.	Bokaro		

Source : Records of the COMPFED.

In 1985 COMPFED took over all the sick dairy and chilling plants from the Government of Bihar on the basis of management and also undertook massive repairs and renovation of all these sick dairy and plants. Initially the progress of COMPFED for repairing and renovating the plants was very slow because it required lot of efforts in winning over confidence of farmers. Milk production and enhancement inputs were introduced by the COMPFED and its unions. These inputs are in the form of animal health coverage, artificial insemination, feed, fodders and training. In winning the confidence of farmers of Bihar and improving productivity of animals regular supply of milk production and enhancement inputs has been main instrumental.[19]

In Bihar including, COMPFED has around 30 small and big plants covering 23 districts of Bihar. In this state milk procured by the unions is standardised, processed packaged and/or converted into products, and sold under the brand name of Sudha.

COMPFED and its Unions manufacture following five type milks :

(a) Full Cream Milk (Sudha Gold).
(b) Standard Milk (Sudha Shakti).
(c) Toned Milk (Sudha Healthee).
(d) Double Tonned Milk (Sudha Smart).
(e) Skim Milk (Sudha Lite).

Besides, liquid milk in starch product line includes ghee, table butter, ice-cream, peda, lassi, misti dahi, paneer, kalakand, gulabjamun, rasogulla, milk cake, etc. These days in milk and milk product Sudha is household name. Sudha under COMPFED is the first in the country which has introduced Sugar-free Ice-cream aiming at diabetes and health conscious people. COMPFED's most significant contribution has been the difference it has made in the lives of women in Bihar where gender discrimination has had a long history. All the milks manufactured by COMPFED have become quite popular in Bihar.

PROGRESS OF COMPFED DURING POST-ECONOMIC REFORMS

COMPFED has made a remarkable progress since 1991[20] particularly during post-economic reforms period. At the end of Eighth Plan these were as many as 2127 functional DCCs with a membership of about 1.38 lakh. As noted above there are 5 Milk Unions affiliated to COMPFED as also five dairies under the direct control of COMPFED in Bihar. Facilities in terms of processing cattle feed and training in these units are provided in following table

In addition to the central training centre at COMPFED's headquarters Barauni, Mithila, Tirhut and Vaishali Patliputra have their own training centres. This table shows the training and cattle feed facilities at various milksheds during the year

2004 particularly in Barauni, Mithila, Shahabad, Bokaro, Gaya, Bhagalpur, Ranchi, Jamshedpur and COMPFED containing per day milk in thousand liters per day dairying in tonnes chilling per day in thousand liters, chilling and training centres and cattle feed per day in thousand in Bihar and Jharkhand.

Further, there are two powder plants at Barauni and Vaishali Patliputra. The total installed capacity of milk processing for the federation is around 7.86 lakh liters/day. There are two cattle plants of 100 tonnes each at Patna and Ranchi. Progress of the COMPFED includes achievements of all these unions/dairies. Indicators like DCSs organised, registered and functional, membership of both men and women, procurement of milk. DCSs covered under AIs. AIs carried, sale of cattle feed, milk marketing and vaccinations done are provided in Table 7.15.

TABLE 7.15

Milk Processing* Training and Cattle Feed Facilities at Various Milksheds, 2004

Sl. No.	*Milksheds*	*Milk (000 litres/ day)*	*Drying (tones/ day)*	*Chilling (000 litres/ day)*	*Chilling centres*	*Training centres*	*Cattle feed (tones/ day)*
1.	Barauni	1.80	5.0	0.50	1	1	—
2.	Mithila	0.50	—	0.50	2	1	—
3.	Shahabad	0.70	—	—	—	—	—
4.	Tirhut	0.70	—	0.23	2	1	—
5.	Vishali Patliputra	1.60	5.0	0.57	3	1	100**
6.	Ranchi	0.70					100**
7.	Jamshedpur	1.10					
8.	Bokaro	0.35					
9.	Gaya	0.16					
10.	Bhagalpur	0.35					
11.	COMPFED	7.86	4.0	1.91	5	8	200

* There are eight chilling centres owned by the Government of Bihar and managed by COMPFED.

** Union's feed plant.

*** Owned by Government but managed by COMPFED.

Source : Records of the COMPFED, Patna.

DCSs organised have more than doubled during 1991-2004. There are as many as 4018 DCS organised at the end of March 2004 of which 3285 (82%) are functional while 1647 (41%) are registered. Total membership of these DCSs stood at 2.14 lakh in 2003-04 of which 30909 (14.4%) registered.

As against an installed capacity of 6.86 lakh/days average procurement of federation is 3.42 (50%) lakh in 2001-02. Of the 3285 functional DCSs as many as 2598 are covered under AIs and federation perform 5.36 lakh insemination during 2002-03. As of cattle feed, COMPFED sold 33464 tonnes in 2002-03 selling 268 gram/litre of milk collected. Growth in functional DCSs average procurement of milk/member and sale of cattle feed/kg. milk procured by the COMPFED during 1991-2003 are provided in Tables 7.16 and 7.17.

The major products of COMPFED are butter, ghee, peda, dahi and lassi followed by growth in share capital turnover and net loss/profit in following Tables 7.18 and 7.19 Share capital increased from Rs. 3.95 crore in 1990-91 to Rs. 7.10 crores in 2001-02. Turnover of the federation was Rs. 300 crores in 2001-02 and incurred net losses 5 out of 13 years considered. It has been posting profits continuously from 1996-97. Since the inception of COMPFED impressive advances have been made in mechanised production of a wide range of indigenous milk products. That large scale production involves intensive engineering inputs for designing the plant layout, selection of various equipment and manufacturing processes.[21] Traditional milk products are part of the Indian heritage, with their origin going back to the dawn of Indian civilization.[22] They are today part and parcel of social life, festivals and ceremonies in which ghee and sweets play an important role. In response to the changing life style and emerging consumer demand for food, safety and extended shelf-life, the traditional method of making Indian milk products in Bihar also, are undergoing major transformation with the introduction of technology.

COMPFED covers capital investment on land, building, equipment, infrastructure, working capital, manufacturing and marketing cost, and financial projection to work out the viability of the plants.

Table 7.16
Progress of COMPFED, 1991-2003

Sl. No.	Indicators	1990-91	1991-92	1992-93	1993-94	1994-95	1995-96	1996-97	1997-98	1998-99	1999-2000	2000-01	2001-02	2002-03
	(1)	(2)	(3)	(4)	(5)	(6)	(7)	(8)	(9)	(10)	(11)	(12)	(13)	(14)
1.	DCSs organised	1967	2041	2008	2288	2417	2633	2728	2844	3002	3371	3525	3792	4018
2.	DCSs registered	1151	1175	1258	1360	1364	1480	3475	1456	1413	1605	1659	1719	1647
3.	Functional	1642	1573	1609	1744	1781	1975	2127	2306	2523	2868	2890	3088	3285
4.	Membership (lakhs)	0.98	1.04	1.02	1.14	1.18	1.32	1.38	1.48	1.58	1.73	1.84	2.00	2,14
5.	Men	0.88	0.94	0.91	0.10	1.03	1.14	1.19	1.26	1.36	1.50	1.59	1.72	1.83
6.	Women	0.09	0.10	0.11	0.14	0.15	0.18	0.19	0.21	0.21	0.23	0.25	0.28	0.31
7.	Procurement (000 kg/day)	94.90	86.89	105.53	119.13	114.0	157.4	195.8	213.2	194.1	281.1	329.8	342.8	384.5
8.	Milk supplied/ membership Kg.	353	305	378	382	353	436	517	527	450	593	654	624	657
9.	DCSs covered under AIS	329	391	552	740	989	1189	1573	1650	1967	2110	2385	2608	2679
10.	AIS done (lakh)	0.74	1.01	1.01	1.08	1.52	1.90	2.49	2.95	3.52	3.79	4.37	4.77	5.37
11.	Cattle feed sold (tonnes)	4151	5665	7512	11288	10017	12039	18611	24944	27378	32607	22760	26554	30566
12.	Sale of cattle feed/kg. of milk procured (g.)	120	179	195	260	241	210	260	321	386	318	189	213	218
13.	Milk marketing (000/lit./day)	—	107.96	106.54	126.56	165.82	179.96	210.36	245.02	303.91	316.97	326.71	331.71	378.63
14.	Vaccinations done (lakh)					1.77	1.87	2.30	2.71	3.10	3.50	4.10	3.27	4.75

Source : Records of the COMPFED. 2004.

TABLE 7.17

Growth and Procurement of Milk and Functions : DCSs under COMPFED

Sl. No. Contents	1990-91	1991-92	1992-93	1993-94	1994-95	1995-96	1996-97	1997-98	1998-99	1999-2000	2000-01	2001-02	2002-03
(1)	(2)	(3)	(4)	(5)	(6)	(7)	(8)	(9)	(10)	(11)	(12)	(13)	(14)
1. Procurement (000 kg./day)	353	305	378	382	354	440	517	527	450	593	655	624	656
2. Growth of Average Procurement of Milk/Member Dairy 1991-03	120	119	195	260	241	210	260	321	386	318	189	213	218
3. Functional	1642	1573	1609	1744	1781	1975	2127	2306	2523	2868	2890	3088	3285

Source : *Ibid.*

For products manufacture, procurement of good quality, raw milk in desired quantity should be assured. The milk collection has to be in tune with the products, manufacture schedule. However, some products can be manufactured in the flush season to be marketed in the lean season.[23]

Strategies adopted for the turn around of these unions are provided in it.[24] Improved efficiency of plants by reducing breakdowns, increased milk procurement and marketing, production and marketing of value added products, and cost control techniques contributed to improved health.

As obvious from Table 7.18 ghee, butter, peda, dahi and lassi are the major product of COMPFED. Table 7.19 reveals the share capital turnover and net profit of COMPFED during post-reform period, i.e. 1991-2004. Both the tables show the trend of progress of COMPFED during 1990-91 to 2002-03.

TABLE 7.18
Growth in Production of Milk Products

Year	*Ghee (Tonnes)*	*Butter (Tonnes)*	*Peda (Tonnes)*	*Dahi (Tonnes)*	*Lassi (000 Lts.)*
1990-91	422	NA	NA	NA	NA
1991-92	393	NA	NA	NA	NA
1992-93	718	NA	NA	NA	NA
1993-94	581	21*	10*	NA	NA
1994-95	444	32.80	69.90	30.79	79.50
1995-96	506	55.14	115.50	81.53	237.40
1996-97	650	67.23	176.00	166.60	384.90
1997-98	642	108.80	205.10	157.60	333.20
1998-99	484	92.40	268.00	337.9	1171.70
1999-00	642	82.25	333.70	401.70	2223.60
2000-01	683	72.70	359.20	439.60	1111.00
2001-02	998	67.30	350.50	387.60	1271.10
2002-03	1749	81.90	407.30	492.60	1070.9

* Starting year.

Source : Record of the COMPFED.

TABLE 7.19

Growth in the Capital, Turnover and Net Profit of COMPFED, 1991-2003

Year	*Share Capital (Rs. Lakhs)*	*Turnover (Rs. Crores)*	*Net Profit (Rs. Lakhs)*
1990-91	494.82	NA	- 335.00
1991-92	494.82	NA	–143.69
1992-93	494.82	NA	–174.58
1993-94	494.82	NA	–25.05
1994-95	494.82	83.10	4.70
1995-96	648.82	117.07	–137.00
1996-97	648.82	159.15	36.60
1997-98	648.82	191.62	109.20
1998-99	649.32	239.81	344.30
1999-00	657.94	280.47	330.90
2000-01	677.45	281.90	690.60
2001-02	709.86	299.61	466.40
2002-03	861.89	340.77	526.28

Source : Record of the COMPFED.

Ghee production has rapidly increased from 422 tonnes in 1990-91 to 1750 tonnes in 2002-03 more than four times high, similarly production of butter has also enhanced from 21 in 1993-94 tonnes to 82 tonnes during the period 2002-03. The peda production was only 10 tonnes in 1993-94 but very rapidly increased to 340.77 tonnes in 2003-04 which was more than 40 tonnes high. The production of dahi was about 40 tonnes in 1994-95 which increased to 493 tonnes about 17 times high and lassi production also increased from 80 thousand liters to 1071 thousand liters. In other words, we find rapid growth of milk products during the period 1991-92 to 2002-03 of COMFED in Bihar. Table 7.19 shows that the share capital of COMFED has increased Rs. 495 lakh in 1991-92 to Rs. 862 lakh in 2002-03 in turnover increased from Rs. 83.10 lakh in 1994-95 to 340.77 in 2002-03. Till 1995-96 Bihar State Dairy Development was incurring heavy losses but since 1966-67 COMPFED has been incurring profits. The profit of COMPFED in Bihar was about

Rs. 37 lakh which increased to Rs. 691 lakh in 2000-01. But its profit reduced to Rs. 466.40 lakh in 2001-02. However, its profit improved further to Rs. 526.26 lakh in 2002-03.

PROGRESS OF BARAUNI DAIRY

Barauni Dairy is managed and commanded by Desh Ratna Rajendra Prasad Dugdh Utpadak Sahkari Sangh Limited (Barauni union). The areas come under the command of Barauni Dairy are Barauni, Begusarai and Khagaria. The Desh Ratna Rajendra Prasad Dugdh Utpadak Sahkari Sangh known as Barauni union was registered in 1987 which was in 1992 transferred to an elected Board. The Board comprises 16 Directors out of which 12 are elected. There is also a provision for women directors on the Board. The Barauni dairy union has a processing plant with an installed capacity of 1.6 lakh liters a day, a chilling plant (4000 litres/days) and a bulk cooler (5000 litres/day), Ghee, Peda, Kalakand, Rasogulla, Gulabjamun, Misti Dahi, Plain Dahi, Lassi, Paneer, Khoa, Milk cake, Dairy whitener, WMP, Khoa Mithai. Sudha Energy Bar Butter, Paneer Bhunjia are some of the products being manufactured at the Barauni dairy. Barauni dairy has very limited market for liquid milk. Its milk is sent to Ranchi, Patna, Kolkata, Jamshedpur, Bokaro, Bhagalpur, Gaya, Biharsharif, Barh, etc.

Table 7.20 deals with the progress of Barauni dairy between 1991 to 2003. It shows that profile of COMPFED particularly progress of Barauni dairy and union comprising dairy co-operative societies registered, Dairy Co-operative Societies functioning, membership of DCSs with men and women, procurement of milk per day (thousand kg.), payment to producers (in lakhs of Rupees), cost of milk in (Rs. per litre). DCSs covered under AI, AIs performed, AIs lady workers cattle feed sold in tonnes per annum, DCSs having owned buildings, milk marketing in thousand litres per day/people before interest and depreciation in Rs. lakh and profit after tax. Table 7.20 presents the profit of COMPFED Milk union of Barauni dairy. It can be observed from table that number of DCSs registered increased from 242 in 1990-91 to 370 in 2002-03 and DCSs functioning also increased from 396 in 1990-91 to 647

TABLE 7.20
Progress of Barauni Union, 1991-2003

Sl. No.	Contents	1990-91	1991- 92	1992-93	1993-94	1994-95	1995-96	1996-97
	(1)	(2)	(3)	(4)	(5)	(6)	(7)	(8)
1.	DCSs Organised	242	253	289	297	322	330	333
2.	DCSs Functioning	396	417	395	418	456	460	482
3.	**Membership**	22953	23979	24324	27920	29543	34007	35737
	Men	21560	22408	22662	25568	27056	31111	32306
	Women	1393	1571	1662	2352	2487	2896	3431
4.	Procurement (000 kg/ day)	33.54	30.35	34.36	40.61	37.66	52.38	60.51
5.	Capital Utilization (%)	33.54	30.35	34.36	40.61	37.66	52.38	60.51
6.	Payment of Producers (Rs. lakh)			969.45	1154.66	1039.68	1781.86	2095.98
7.	Cost of Milk (Rs. /litre)			7.73	7.79	7.56	9.32	9.49
8.	DCSs Covered under AI	88	72	144	-156	176	374	420
9.	AIS Performed	22282	29704	33966	36963	53081	72322	92539
10.	All Lady Workers	3	2	4	4	4	5	6
11.	Cattle Feed Sold (tonnes)	920.05	1348.67	1402.37	2037.37	1971.32	2366.25	3131.18
12.	DCSs having owned Building							
13.	Milk Marketing (000 ltr / day)					5.37	4.55	3.66
14.	Profit before Interest and depreciation (Rs. lakhs)	8.08			33.19	37.37	165.48	75.27
15.	Profit after Tax	1.26	–17.25	–40.35	–20.91	–20.64	24.70	–32.33

(*Contd.*)

TABLE 7.20 (Contd.)

Sl. No.	Contents	1997-98	1998-99	1999-2000	2000-01	2001-02	2002-03
	(1)	(9)	(10)	(11)	(12)	(13)	(14)
1.	DCSs Organised	329	308	342	363	367	370
2.	DCSs Functioning	533	547	552	574	602	647
3.	Membership	0.35	0.35	0.36	0.41	0.42	0.49
	Men	0.31	0.31	0.32	0.36	0.37	0.43
	Women	0.04	0.04	0.04	0.48	0.06	0.06
4.	Procurement (000 kg./day)	65.93	57.19	71.37	85.21	106.64	134.38
5.	Capital Utilization (%)	65.93	57.19	44.60	53.25	66.65	83.97
6.	Payment of Producers (Rs. lakh)		1891.21	2524.25	2979.54	3600.43	4462.76
7.	Cost of Milk (Rs./litre)	9.06	9.69	9.58	9.25		9.10
8.	DCSs Covered under AI	423	457	484	503	568	599
9.	AIS Performed	105377	107470	115727	125626	131810	134709
10.	All Lady Workers	13	13	11+	17	25	25
11.	Cattle Feed Sold (tonnes)	3228.13	4994.02	4486.85	1953.23	3175.05	3926.45
12.	DCSs having Owned Building						
13.	Milk Marketing (000 ltr./day)	6.85	8.81	7.58	8.62	12.18	17.27
14.	Profit before Interest and Depreciation (Rs. lakhs)	145.60	129.74	46.54	198.69	245.93	230.20
15.	Profit after Tax	5.74	1.61	-98.96			

Source : Records of the COMFED.

in 2002-03, similarly its procurement also rapidly increased from 33.54 thousand per day in 1990-91 to 134.36 thousand per day. Its capacity utilization enhanced from 33.54% in 1990-91 to 84% during 2002-03 and payment to producers went from Rs. 969.45 lakh in 1992-93 to Rs. 4663 lakh in 2002-03. Similarly cattle feed sold during 1991-92 to 2002-03 was 920.5 and 3225.45 thousand tonnes respectively. Profit depreciation increased rapidly from Rs. 8.08 lakh to Rs. 230.20 lakh which is a matter of Satisfaction.

PROGRESS OF RANCHI DAIRY

Ranchi Dairy is an important dairy of Jharkhand but directly controlled under COMPFED located is the capital of Jharkhand Ranchi. Ranchi Dairy is mostly a processing and marketing centre. However, off late, few DCSs are also organised in Ranchi Dairy. The dairy covers Ranchi, Lohardaga and Gumla districts and has an installed capacity of 70000 litres/day. Apart from selling liquid milk at Ranchi, Hazaribagh, Ramgarh, Ghato, Simdiga, Gumla, Muri, Jhalda, etc., the unit manufactures peda, paneer, lassi, misti dahi and fitness curd and has 60 employees. It has organised 111 DCSs of which 78 are functioning (Table). It is to be mentioned here that the command area is drought prone and dominated by tribals. Hence, there is less scope for organizing DCSs and collecting milk from this area. However, dairy has been making efforts to motivate farmers going for dairying.

Average procurement is around 7000 litres/day and the dairy paid Rs. 2.80 crore to producers in 2002-03. Only 20 out of 78 DCSs are covered under AI network particulars on AIs performed. AI lady workers and cattle feed sold are also provided in the Table 7.21.

PROGRESS OF VAISHAL PATLIPUTRA DAIRY AND MILK UNION

Patna was one of the milksheds identified under Operation Flood-I, and a lakh litres/day capacity feeder balancing dairy and 100 tonnes/day capacity feed plant were set-up in 1975.[25] A spear head team was also deputed from *NDDB* in the same year.

TABLE 7.21

Progress of Ranchi Dairy,1991-2003

(In Million numbers)

Sl. No.	*Contents*	*1994-95*	*1995-96*	*1996-97*	*1997-98*	*1998-99*	*1999-2000*	*2000-01*	*2001-02*	*2002-03*	*2003-04*
	(1)	*(2)*	*(3)*	*(4)*	*(5)*	*(6)*	*(7)*	*(8)*	*(9)*	*(10)*	*(11)*
1.	DCSs Registered				5	5	18	23	23	23	24
2.	DCSs Functioning	7	10	20	35	50	65	65	68	72	78
3.	Procurement (000 kg./day)	0.7	1.6	3.3	5.3	6.2	7	5.7	4.9	5.9	6.0
4.	Payment to Producers (Rs. lakh)	3.9	38.2	88.3	171.8	203.7	245.3	189.3	162.7	200.3	200.9
5.	Cost of Milk (Rs./litre)	6.15	6.54	7.33	8.88	9.0	9.6	9.1	9.1	9.3	9.8
6.	DCSs Covered under AI				2	11	16	17	18	19	20
7.	AIS Performed				25	734	1671	2177	2910	3399	3403
8.	All Lady Workers						2	2	3	3	3
9.	Cattle Feed Sold (tonnes)	44	156	279	412	634	571	278	320	238	256
10.	Milk Marketing (000 litres/day)	18.60	23.62	26.27	30.02	34.73	35.27	36.66	41.44	44.17	46.71

Source : Records of the Union.

As the progress under *BSDDC* was not satisfactory, management of dairy development activities were handed over to *NDDB* in 1981. *NDDB* appointed an integrated spear head team to restructure milk procurement and to streamline working of Patna dairy project. With a view to oversee dairying in the state, *NDDB* also took interest in organising *Vaishal Patliputra Dugdh Utpadak Sahakari Sangh Limited* in 1987, and Patna Dairy Project was handed over to the union in 1988. The union covers Patna, Vaishali and Nalanda and fringe areas of Saran districts, and has 12 elected and three nominated Board of Directors. There is a provision for one each woman from schedule caste and schedule tribes. Apart from having a feeder balancing dairy with a capacity of 1.5 lakh litres/day, the union has a chilling plant (40000 litres/day) and two bulk coolers (5000 litres/day each) as demonstrated in Tables 7.22 and 23.

PROGRESS OF SHAHABAD DAIRY AND MILK UNION

Hindu king Shershah Suri who defeated Akbar the Great ruled this part of the country.[26] A chilling centre at Arrah was under the direct control of Vaishal Patliputra milk union for some time. In order to meet the aspirations of local farming community *Shahabad Dugdh Utpadak Sahakari Sangh Limited* was registered in 1997 with ten elected Board of Directors. There is provision for a woman on the board Shahabad milk union covers north Bihar districts of Bhojpur, Buxar, Rohtas and Kaimur. It is a fledgling milk union of affiliated to *COMPFED*. Since inception milk union has been struggling hard due to inadequate funds, poor infrastructure in terms of roads and communication, illiteracy and above all poverty.

The union has two chilling centres at Kochas and Dumarawn with an installed capacity of 30,000 litres/day. Ghee and various grades of milk and manufactured, and sold at Sasaram, Arrah of Bihar, Varanasi, Gazipur and Baliya of U.P. and the union has 72 employees.

Progress of the milk union is provided in Table 7.24. At the end of 2003-03, it has 461 functional societies of which 137 are registered. Total membership of the union is 28000 (2002-03) of which 1025 are women. Substantial improvement in procurement of milk/day can be noticed from the Table 7.24.

TABLE 7.22
Progress of Vaishal Patliputra Union, 1991-2003

Sl. No.	Contents	1990-91	1991- 92	1992-93	1993-94	1994-95	1995-96	1996-97
	(1)	(2)	(3)	(4)	(5)	(6)	(7)	(8)
1.	DCSs Organised	456	461	471	516	582	611	624
2.	DCSs Functioning	706	726	730	737	765	833	878
3.	Membership (lakhs)	0.38	0.42	0.43	0.46	0.48	0.52	0.57
	Men	0.35	0.39	0.40	0.42	0.43	0.46	0.50
	Women	0.03	0.03	0.03	0.04	0.05	0.06	0.07
4.	Procurement (kg./day)	30994	27572	36068	41556	38632	45907	57627
5.	Capital Utilization (%)	40	38	40	35	39	41	51
6.	Payment of Producers (Rs. lakh)	628.04	666.01	868.06	1033.22	1115.61	1570.83	2218.38
7.	Cost of Milk (Rs./litre)	4.53	4.95	5.55	5.96	6.52	7.53	8.03
8.	DCSs Covered under AI	134	124	233	314	394	487	553
9.	AIS Performed	32257	43878	36415	38896	49119	60999	82520
10.	All Lady Workers	4		6	7	10	12	14
11.	Cattle Feed Sold (tonnes)	11446	13557	13294	13944	12884	13973	18827
12.	Milk Marketing (000 ltr./day)							49.55
13.	Profit before Interest and Depreciation (Rs. lakhs)	53.63				101.55	98.88	173.22
14.	Profit after Tax	–19.14	--20.93	–28.25	20.02	7.90	39.01	12.93

(*Contd.*)

TABLE 7.22 (Contd.)

Sl. No.	Contents	1997-98	1998-99	1999-2000	2000-01	2001-02	2002-03
	(1)	(9)	(10)	(11)	(12)	(13)	(14)
1.	DCSs Organised	507	520	610	625	651	651
2.	DCSs Functioning	972	826	880	894	913	1007
3.	Membership (lakhs)	0.63	0.52	0.53	0.56	0.58	0.61
	Men	0.55	0.44	0.45	0.48	0.49	0.52
	Women	0.08	0.08	0.07	0.06	0.09	0.10
4.	Procurement (kg./day)	82342	97312	88382	97312	88382	96754
5.	Capital Utilization (%)	57	62	61	58	59	66
6.	Payment of Producers (Rs. lakh)	2353.15	1912.75	3425.28	3896.8	3283.83	3516.3
7.	Cost of Milk (Rs./litre)	8.86	9.86	11. 36	10.95	10.20	9.95
8.	DCSs Covered under AI	634	651	667	673	687	750
9.	AIS Performed	97042	104125	104607	116426	123266	135299
10.	All Lady Workers	23	28	30	33	36	39
11.	Cattle Feed Sold (tonnes)	21485	22667	26031	18640	21033	23845
12.	Milk Marketing (000 ltr./day)	68.84	80.16	79.65	76.38	76.70	89.77
13.	Profit before Interest and Depreciation (Rs. lakhs)	145.90	169.87	176.60	165.06	154.92	234.40
14.	Profit after Tax	19.11	19.12	10.02			175.29

Source : Records of the Union.

TABLE 7.23
Progress of Milk Union in Vishal Particular Union During 1991-2003

Year	Function of DCs (Numbers)	Procurement Kg./day	Growth of Membership	Growth in Average Product Price (Litre of Milk Rs.)
1990	700	30994	3800	5.58
1991	726	27572	4200	6.62
1992	730	36068	4254	5.48
1993	737	41556	4608	4.62
1994	765	38632	4827	6.28
1995	833	45907	5261	7.35
1996	878	57627	5662	8.0
1997	972	58348	6276	8.75
1998	820	53165	5196	9.66
1999	880	82342	5308	11.36
2000	994	97312	5552	11.36
2001	913	88382	5769	10.99
2002	1997	96754	6130	10.25

Source : Ibid.

The union paid about Rs. 11 crores to producers in 2002-03 which was just Rs. 8.2 crores in 1997-98. Thus, the cost of milk/ litre increased from Rs. 8.81 to Rs. 9.25 during 1998-2003. The union has been very effective in providing inputs like AIs and cattle feed. It can be seen from the table that about 400 out of 461 DCSs are covered under AI network and performed about 90000 AIs.

PROGRESS OF MITHILA DAIRY AND MILK UNION

Mithila Dugdh Utpadak Sahakari Sangh Limited, Samastipur covers Samastipur, Darbhanga and Madhubani district of Bihar. Union was registered in 1987 and transferred to an elected Board in 1989. The Board has 16 Directors of which 12 are

TABLE 7.24

Progress of Shahabad Milk Union, 1997-03

Sl. No.	Contents	1997-98	1998-99	1999-2000	2000-01	2001-02	2002-03
	(1)	(2)	(3)	(4)	(5)	(6)	(7)
1.	DCSs Registered	99	99	128	131	132	137
2.	DCSs Functioning	245	338	407	426	436	461
3.	Membership	14150	18250	28800	24060	27000	28000
	Men	13939	17981	22279	23339	26004	26975
	Women	211	269	521	721	996	1025
4.	Procurement (000 kg./day)	8.77	23.39	35.48	36.65	33.81	32.86
5.	Capacity Utilization (%)	30	30	30	60	60	100
6.	Payment to Producers (Rs. lakhs)	282.06	775.09	1205.79	1217.6	1138.4	1109.7
7.	Cost of Milk (Rs./litre)	8.81	9.08	9.31	9.10	9.22	9.25
8.	DCSs Covered under AI	130	158	216	30	400	398
9.	AIs Performed	16531	21191	33695	53460	72715	88856
10.	AI Lady Workers			3	3		
11.	Cattle Feed Sold (tonnes)	485.6	1820.1	1227.4	670.52	519.91	658.50
12.	Milk Marketing (000 litres/day)	0.04	2.34	3.64	3.66	3.82	12.72
13.	Profit above Interest and Depreciation (Rs. lakhs)	12.05	24.15	83.86	-43.52	-6.10	50.72

Source : Records of the Union.

elected. The union implemented Operation Flood Programme in the command area. Major objectives of the union are:[27]

(a) organisation to DCSs on amul pattern,
(b) arranging milk procurement and providing remunerative prices to producers,
(c) productivity improvement through supply of balanced cattle feed, fodder seeds, and by-pass protein feed, and
(d) arranging training programmes for members and employees working in the DCSs.

The union has chilling centres at Darbhanga and Rosera, and a processing plant at Samastipur. Total installed capacity of the union is 80000 litres/day. The milk so procured from the DCSs is sold at Samastipur, Darbhanga and Madhubani towns. It has 160 employees on its roll.

Progress of the union for the period 1994-2003 is providing in Table 7.25. Union has 639 DCSs of which 88% are functional with a membership of 367.46. Of the total membership, about 19% are women. There is substantial improvement in membership and DCSs organised during study period. So is the case with procurement which steadily increased from 13000 litres/day in 1993-94 to 62550 litres/day in 2002-03. As a result, capacity utilisation of the union has increased substantially to 78%. The union has a good AI network and performed 1.07 lakhs AIs through 423 AI centres in 2002-03. There are 41 lady workers too undertaking AI activities in the union. The union sold as much as 4029 tonnes of cattle feed thereby selling 176 g. of feed for every kg. of milk procured. About 18 DCSs have their own buildings. Details on milk marketing as also financial health are also provided in the Table 7.25.

PROGRESS OF TIRHUT DAIRY AND MILK UNION

Muzaffarpur dairy with a capacity of 25000 litres/day was established in 1982 by Government of Bihar with financial aid from *'Freedom from Hunger Campaign'*, a UK-based voluntary organization. The project was aimed at meeting demand for liquid milk in various urban towns of North Bihar and to improve economic conditions of rural milk producers. In due course, milk procurement and marketing activity was initiated by BSDDC which controlled the plant till April 1984. However, procurement was less than 1000 litres/day during this period.

The plant was handed over to COMPFED in April 1984 to implement Operation Flood Programme. Subsequently, Tirhut Dugdh Utpadak Sahakari Sangh Limited was registered in 1988 having Muzaffarpur, Sitamarhi and East Champaran as its area of operation. Objective of the union is to generate and/or supplement income of milk producers. Highly flood prone command area and skewed distribution of land has caused an alarming increase in unemployment among rural poor. The milk plant started procurement on June 1, 1984[28] followed by marketing in July 10th of the same year. With a view to accelerate rural income and employment, union has adopted modern technology to improve productivity. Entire activity of procurement, technical inputs, processing and marketing of milk and milk products of Muzaffarpur dairy/Muzaffarpur milkshed was handed over to Tirhut milk union in January 1991. Plant capacity was increased to 60000 litres/day in 1995 and further to 1 lakh litres/day in 2000. Chilling centres located at Sitamarhi, Motihari, Bettiah, Gopalganj and Sahebganj.

Progress of the Union

Data on progress of milk union are provided in Table 7.25. At the outset, a spectacular progress can be noticed during the last 13 years for which data are available. Currently, there are 1007 functional DCSs with a membership of 61307. There are as many as 9626 women members in about 160 WDCSs. Daily average procurement increased from 30994 kg/day to 96754 kg/day during 1991-2003. While the union has fairly good number of functional societies, emphasis is being given to consolidate functioning of DCSs by increasing members' participation. Capacity utilisation of the union is around 60% which increased from around 40% in the early 1990s. The union paid an amount of Rs. 35.16 crores to producers in 2002-03.

In addition to providing ready and stable market for rurally produced milk, union has been providing milk production enhancement inputs like artificial insemination with frozen semen, veterinary first aid, vaccination, supply of

TABLE 7.25
Progress of Mithila Milk Union, 1994-2003

(*In Million numbers*)

Sl. No.	*Contents*	*1993-94*	*1994-95*	*1995-96*	*1996-97*	*1997-98*	*1998-99*	*1999-2000*	*2000-01*	*2001-02*	*2002-03*
	(1)	*(2)*	*(3)*	*(4)*	*(5)*	*(6)*	*(7)*	*(8)*	*(9)*	*(10)*	*(11)*
1.	DCSs Registered	333	393	416	393	428	493	566	607	665	639
2.	DCSs Functioning	256	227	271	322	373	472	520	498	570	564
3.	Membership	14869	14932	18034	17213	19800	24850	30740	34930	37529	36746
	Men	11667	11509	14123	12947	14872	19330	24510	28293	30372	29936
	Women	3202	3423	3911	4266	4928	5520	6230	6637	7157	6810
4.	Procurement (000 kg./day)	13.48	10.40	16.94	26.68	31.99	30.35	42.55	45.04	56.41	62.55
5.	Capacity Utilization (%)	22.48	17.33	28.23	44.46	40	37.93	53.19	67.55	70.51	78.19
6.	Cost of Milk (Rs./litre)	5.00	5.70	6.25	7.00	7.40	8.50	9.00	8.75	8.63	8.00
7.	DCSs Covered under AI	162	193	226	246	278	380	394	434	492	423
8.	AIs Performed	21941	33420	37544	56704	66821	81664	94529	96975	99862	10743
9.	AI Lady Workers	4	5	14	22	27	28	30	32	41	41
10.	Cattle Feed Sold (tonnes)	1064.42	1008.75	1499.33	3176.75	4987.5	5437.6	6033.63	3254.58	4015.58	4029.12
11.	DCSs having Owned Building	9		1	1	5	1	1			18
12.	Milk Marketing (000 litres/day)		10.13	9.95	15.27	21.55	31.37	37.00	41.92	41.45	42.99
13.	Profit before Interest and Depreciation (Rs. lakhs)	-0.77	2.41	28.46	39.48	35.83	59.05	69.40	153.23	128.07	119.38
14.	Profit after Tax (Rs. lakhs)	-10.56	-10.62	4.09	6.61	1.74	-15.78	-20.48			0.87

Source : Records of the Union.

balanced feed, supply of fodder seeds, treatment of paddy straw/wheat *bhusa* with urea, supply of urea molasses block, etc. About 750 DCSs are covered under AI facility and performed 1.35 lakhs of AIs during 2002-03. There are as many as 39 AI lady workers of which only 20 are working at the moment. Union sold 23485 tonnes of cattle feed.

It has facilities to manufacture milk powder, butter, ghee, ice cream, peda, paneer and plant/misti dahi. Union introduced butter in October, 1993, ice cream in April 1995, and plain mist dahi in November 2001. Data on functional DCSs, membership, average/day milk procurement average producers price/litre of milk, cattle feed consumption/kg of milk procured and production of ghee and powder are provided.

Clean Milk Production

In tune with the changes in general environment, union responded favourably, for producing quality milk and milk products. Besides creating awareness amongst milk producers about importance of clean milk production, efforts were made to install bulk coolers. Following initiatives too were taken in this regard :

(a) pre-poning arrival time of milk vehicles at the chilling centre/dairy docks,
(b) use of SS-304 milk cans replacing the aluminium cans. Use of stainless steel milk testing equipments were introduced and distribution of stainless steel vessels to milk producers are bonus was also encouraged,
(c) cleaning and sanitization of milk cans,
(d) testing adulterants in milk with kits provided by NDDB, and
(e) covering milk procurement vehicles to protect milk cans from exposure to direct sun-light.

Productivity and Quality

Dairy plant management programme (DPMP) was introduced in 1992 followed by quality assurance programme (QAP) in 1993. This resulted in bringing about a positive change leading to viability coupled with decreasing operational costs on the one hand, and improved quality of products on the other. ISO certification process was completed leading to quality management system and food safety which lead to ISO 9001 : 2000 and HACCP (IS 15000) in 2002.

Future Challenges and Task Ahead

Future areas of the union include making Sudha brand, a market leader, consolidation of existing DCSs leading to increased milk procurement, improvement in active involvement and participation members, producers, bridging the flush lean gap, popularising all technical inputs, increasing throughputs and sale of both milk and milk products as well as cattle feed, reducing handling losses and increasing utilisation of plant capacities, optimising utilisation of all consumable and above all human resource development through training.

Relative Progress during 1991-2003

The union has 492 functional societies of which 260 are registered. There are over 34000 members in these societies of which 54% are poorer members. Women form 26% of the total membership. Union is collecting 44000 litres of milk/day with a peak procurement of 60755 litres/day. Half of the procurement comes from buffalos. Capacity utilization of the union has been more than 100% for the last 10 years, and paying amounts varying between Rs. 1.51 crores to Rs. 15.83 crores during 1991-2003. Substantial improvement in the cost of milk can be noticed during period under review. Input services like AIs and cattle feed are being provided to producers. It has 63 single and 49 cluster AI centres. Further, union supplied 50.50 tonnes of fodder seeds and de-wormed 67105 animals in 2002-03. Union started earning profits from 1993-94, however, profits after tax negative till 1999-2000. The paid up share capital of the union is Rs. 63.10 lakhs with a reserve fund of Rs. 44.89 lakhs (2002). Total turnover of the union was to the tune of Rs. 37.57 crores with a net profit of Rs. 93.95 lakhs.

Profile of Milk Product

In addition to selling milk, union started manufacturing peda in 1993. Present product list include peda (100g. and 200 g. pack), sweet lassi (200 ml pouch), misti dahi (100 g. pack), Surbhi (100 g. pack), Kalakand (100 g. and 200 g. pack), AGMARK table butter (500 g. and 50 g. pack), AGMARK ghee (500 g., 200 g., and 15 kg. jar pack), vacuum pack paneer (200 g. pack), ledikeni (100 g., 250 g., 500 g. and 10 kg. jar), balusahi (125 g., 250 g. and 400 g.), chocoburfi (100 g.) and salted lassi (200 ml. pouch).

Awards to the Union

Excellent work carried out by the union was recognised by a number of agencies and received following awards. National award for dairy development and production in 1994-95 and 1997-98 and Best productivity in 1988-89 from National Productivity Council, New Delhi. The union also received award for plant utilization, consumption of polyfilm, consumption of coal and milk processing from COMPFED in 1997-98, and ISO 9001 : 2000 and HACCAP : 15000.

TRAINING AND MANPOWER DEVELOPMENT

Capacity building/skill up-gradation has been given maximum emphasis in implementing dairy development programmes in the state. This has been achieved through regular training of milk producers, management committee members (MCM) of DCSs, staff of DCSs, milk unions and official of COMPFED. They are trained at Patna as also sent to other milk unions in the country VMNICM, and NDDB. Training programmes organized at COMPFED training centre include society operation for secretaries, orientation of MCM, AI and VFA, dairy animal management, legal literacy and women empowerment, refresher course and tailor made programmes as per the need of milk unions. Training in artificial insemination was started at Patna in 1999-2000.

COMPFED ploughs back around 65-70% of total turnover very year to milk producers which comes to around Rs. 220 crores during 2002-03. It gives 65% of its earned profit to farmers as bonus, dividend and development fund, while 35% is used for plant and machinery maintenance. COMPFED has entered into fruit and vegetables business and exported litchi. However, the experience was not favourable. COMPFED received awards from National Productivity Council and National Commission for Women, and international food quality certification.

INVISIBLE CONTRIBUTION OF COMPFED

Besides providing self-employment to 2.14 lakh farmers, COMPFED has been able to make an impact on the prevailing social system. Producer members from all castes and creeds come together to a common platform, DCS, to pour milk and all of them have equal voting rights. Further, COMPFED and its union implemented other of socio-economic programmes, namely, rural family health project with support from centre for development and population activities of USA, programme of renewable energy sources with support from Bihar Rural Energy Development Agency (BREDA), programme of women empowerment with support from the department of women and child, the ministry of human resource development, and programme of poverty alleviation for the SCs and OBCs with support from national finance and development corporation for SC and OBC.

RELATIVE SHARES OF MILK UNIONS

Based on the descriptions provided on individual milk union, an attempts is being made to ascertain relative shares of each milk union on important parameters like functional DCSs, milk procurement, DCSs covered under AI network and sale of cattle feed. Of the total DCSs of the COMPFED. Vaishal Patliputra has 31% followed by Barauni (20.0%), Mithila (17.4%), Tirhut (15.2%), Shahabad (14.02%) and Ranchi (2.3%). With 20% of DCSs interestingly, Barauni union contributed 36% to total milk procurement as against 26% from Vaishal Patliputra. Thus, procurement/DCSs is the highest in Barauni and the lowest in Ranchi while Mithila stood second (111 litres/

DCS/day). As of AI centres, 29% are concentrated in Vaishal Patliputra followed by Barauni (23.1%) and Mithila (16.3%). In terms of sale of cattle feed, Vaishal Patliputra stood first (61.2%) while Ranchi (1.6%) sold lowest.

COMPFED was conceived to replicate and oversee Anand pattern dairy cooperatives in the State. Federation was lucky enough to get excellent managing directors and dedicated and sincere employees. Federation has drawn as many as 55-60% officers from other states in the formative years. With the support received from NDDB, Government of Bihar and various other international agencies, COMPFED slowly but steadily improved its performance over years.

It has five affiliated milk unions and as many dairies under its direct control covering 23 districts of the state. Total installed capacity of the federation is 6.86 lakh litres/day. COMPFED has over 4000 DCSs with a membership of 2.14 lakhs. At the aggregate level, about 50% of the installed capacities are utilized. Apart from providing number of technical inputs, COMPFED and its unions have been manufacturing number of dairy products are marketed under the brand name of Sudha, which is a family name in the state. The federation also recognised importance of quality and 9 out of 10 dairies have already been accredited with ISO 9000 : 2000 and HACCP (ISO 15000 : 1998). Of the total DCSs of COMFED, Vaishal Patliputra has the highest number and sold 61% of total cattle feed. Barauni has the highest procurement/DCS. Financial health of both COMPFED and unions has been satisfactory.

Thus, in a depressing scenario where almost all state promoted corporations are helplessly swimming in an ocean of red, COMPFED is the only glittering star on the industrial horizon of Bihar. Currently, COMPFED is handling about 10% of the marketable surplus leaving scope for organising the programme in uncovered villages.

Thus, in Bihar also milk is valued commercially for its two important parameters : (a) Milk Fat (F), and (b) Solid Net Fat (SNF). The SNF largely consists of proteins, lactose and minerals. These solids are also referred to as 'serum solids'. These two parameters usually form the basis of payment to milk producers in India. The term 'total solids' (TS) refers to the quantity of SNF plus fat present in milk. It may range to 12 to 16% depending on its sources. For cow milk TS is 12% (3.5% F + 8.5% SNF) while buffalo milk is ranges between 15 and 16% (6-7% F and 9% SNF). Apart from breed related differences, certain other factors also influence the gross composition of milk which are :

(i) individually of animal;
(ii) stage of milking;
(iii) intervals of milking;
(iv) completeness of milking;
(v) frequency of milking;
(vi) irregularity of milking;
(vii) portion of milking;
(viii) different quantens of udder;
(ix) location period;
(x) yield of milk;
(xi) season;
(xii) feed;
(xiii) nutritional level;
(xiv) health status;
(xv) ege;
(xvi) weather;
(xvii) oestrum or heat;
(xviii) gestation period;
(xix) exercise;
(xx) excitement, and
(xxi) administration of drug and hormones.

Moreever, the people's participation and the role of grassroots institutions, e.g. Panchayati Raj Institutions (PRIs), Non-Governmental Organisations (NGOs), Self Help Groups (SHGs) and other social institutions, have become hall mark of dairy development in present globalised world particularly in the case of COMPFED. People's participation, empowerment and good governance are powerful tools that can give poor people a voice in the affairs of the state and government.

Notes and References

1. Rao, V.M. (2005) : *Co-operatives and Dairy Development*, Mittal Publications, New Delhi, p. 9
2. Mishra, S. : *How Women's Health can be Fostered by Dairy Co-operative Societies*. The Story of Change in Samastipur District of Bihar, Patna.
3. Thomes Phillipse (1978): "No White Revoluation without Green Revolution, Yojana, January.
4. White Report 2006 : Govt of Bihar, March 2006.
5. Census 2001 : Registrar General of India.
6. Sen, Amartya and Jeam Jreze (1996) Indian Development.
7. World Bank, 1992.
8. Annual Report (2000) : State Enterprises of Bihar.
9. Hand Book of Animal Husbandary—Publication and Information Development, Dairying, 708, (Bihar).
10. *The Searchlight* (1972) : Patna, 14 March, 1972.
11. *The Gazetter of India*, Vol. III, Agriculture and Allied Industries, p. 367.
12. Rao, S.B.N. (1982) : "Progress of Dairy Industry in Andhra Pradesh", *Southern Economics* 21(14) 35.
13. *Patna Dairy Project* : Annual Report, 1986-86, p. 2.
14. Patric Joh (1975) : Economics of Dairy Development in India, Patna, Prabhat Prakashan, Digheghat.
15. *The Indian Natoon*, 15 July 1988, Patna.
16. Rao Chaudhary, P.N. (1995)
17. COMPFED (2003) : Annual Report, 2003-04, Patna.
18. Rao, V.M. (2003) : Evaluation of Women Dairy Project in Rajasthan: Regency Publication, New Delhi.
19. COMPFED, (2000) : Annual Report, 2000-01, Patna.
20. Conference Volume 2001, Economics Association of Bihar.
21. John Patricks (1970) : "A Case of Selling Upsmell Dairy Farm", *The Sherghligh*, August 30, Patna.
22. John Patricks (1966) : "दुध अमृत है, इसे बर्बाद ना करो। बिहार समाचार, अप्रैल 16", Patna.
23. Singh, Sher (1972) : "Milk Strategy", *Kurukshetra*, April.
24. Thomas Philipse (1978): "Planning for more Milk", *Kurukshetra*, May, Government of India.
25. Rao, K.M. (1980): Dairy Industry in Andhra Pradesh : *Dairying in India*, (1980) IDSA, New Delhi.
26. Singh, R.P. and J.K. Singh (1994) : "Dynamics of Dairy Development in Bihar", *Indian Dairyman*.
27. BSCMPF, Annual Report of Bihar State Co-Operation Milk Products Federation.
28. *The Searchlight*, Patna, 3 June, 1984.
29. *The Indian Natoon*, Patna 15, June 1984.

8

White Revolution in the 21st Century

This book on dairy development has been written with a view to make a micro-analysis and bird's eyes observation of various issues related to White Revolution.

The history of Indian dairy milk products is perhaps as old as Indian cultivation itself. Even our ancestors began to domesticate milk animals, they found innovative ways to convert highly perishable milk into more stable longer lasting milk products. It is a part of Indian culture to never cows, and kings of *yora* often gifted cattle as rewards to their kinsmen. Thus, it is not surprising that Indians have a deep-rooted tradition of using milk and milk products. It is a customary practice to grace the Indian ceremonies and functions with ghee, butter and sweets made from mawa (khoa), chhana and chakka." as stated by (Dr. V. Kurien, the founding father of white Revolution.

Traditional milk products represent the most public segment of our Indian Dairy Industry. The dairy sector has been playing a significant role in improving the economic condition of our rural population. A majority of the rural households of the country own livestocks. Our farmers have benefitted from different dairy development schemes and India today has the distinction of being the highest producer of milk in the world with an average growth rate of 4-6 percent per annum, the credit which largely goes to 10 million farmers associated with dairy farming through more than 82,000 village dairy co-operative societies.

Preparation and marketing of milk are confirmed to the unorganised sector despite the immensity of volume of milk.

Since most of the western type dairy products manufactured by the organised sector of the dairy industry are reaching near saturation level in the existing domestic and international markets, the entire ranging Indian milk products represent the most promising venue for diversification.

Furthermore, dairying has played a prominent role in strengthening our rural economy. It has been recognised as an instrument to bring about socio-economic transformation by helping the landless and marginal farmers as asserted in the Mid Term Appraisal of the Tenth Plan, Eleventh Plan and Union Budget, 2006-07 and 2007-08.

Access to modern technologies, research development and creation of infrastructure facilities has given tremendous boost to dairy development specially in the co-operative sector according to Economic Survey, 2005-06, 2006-07.

When we say 'Indian milk products' we tend to distinguish such products from western milk products such as chees, yogurt, ice-cream, sweetened condensed milk and butter oil.

However, we do have parallels for such western products in the form of paneer, curd/lassi, kulfi, rabri and ghee. Perhaps, the only major western product where it is difficult to draw comparisons is milk powder. In the days following Independence, we have generally concentrated our efforts in encouraging milk production through marketing of liquid milk, either as fresh milk or with the help of recombining milk powder and white butter. This was done for three reasons; *firstly*, to encourage consumption of liquid milk since it provides nutrition in a move wholesome manner as compared to milk products; *secondly*, as compared to value added milk products, liquid milk has to always remained for more affordable; *thirdly*, liquid milk has always been in demand for use as a whitener with tea and coffee.

Over the years, efforts at expanding liquid milk availability through increased milk production have resulted in per capita availability of liquid milk growing from 107 gms. per day in 1970 to a current level of more than 200 gms. per day.

The opportunity provided by increased availability of liquid milk can now be used for efficiently manufacturing and marketing Indian milk products with long shelf-life. This will help in tapping the potential demand for Indian milk products in both the domestic and global markets.

For sustaining further development, nation's dairy industry would have to cope with the rapid transformations that are taking place in the world economies, consequent to GATT agreement. Global trade is being strongly regulated by the WTO guidelines. Newer and stricter sanitary and phytosanitary standards are being formed for regulating quality parameters of the export products. Under these newly emerging circumstances, quality standards for production and processing milk can't remain at variance with the international standard.

The superior quality of dairy products coupled with concerns for environment and product safety will require significant changes in the way milk products are produced and packed.

The needs of the market will determine the change in technology that will be required in future. For changes in socio-economic environment will drive the requirements for traditional dairy products in the newly emerged world scenario.

It is high timely and very strategic to undertake this research project as a catalyst in milk and we should now go further high in the value addition to milk.

All these things could happen with a quality assurance of raw material and technical human resource training at different levels.

This can only be achieved by an appropriate appraisal and awareness to brought on a very large scale since this is a highly dissipated industry and has its origin from tiny, cottage all the way to a large scale industry.

Although, the process of marketing sweets have undergone on continuous change, the time has now come to integrate traditional methods will modern culinary technology to meet consumer demands for better standardized quality, longer shelf like and greater convenience.

A socially responsive approach and more purposeful application of scientific and industrial techniques is required to rebuild our age old practices to ensure the manufacture of indigenous milk products of uniformity high quality.

Thus, Indian dairy products have not only served as a cultural link with the modern dairy industry but also provide a technological based for diversification, export promotion and as a value added product to make the modern dairy industry economically strong to enable the milk producer to benefit from it.

While White Revolution in India/Indian Dairy products have been a source of gray for ages, much need to be done to give their significance they deserve in the global market place.

Growing incomes and consumption have also stimulated not only a greater demand for these product but also provided a unique opportunity for their systematic study, research and documentation relating to their origin, technology and marketing.

Today, Indian Milk Products are the largest and fastest growing segment of white revolution. They offer opportunity for absorbing the growing milk surplus, generated by the operation flood.

WHITE REVOLUTION IN 21ST CENTURY

The White Revolution sector is in search of initiative to enlarge its market in the new millennium. One factor that has lent urgency is the crisis created by continuing decline in world dairy commodity prices. This has occurred despite no significant change in the demand and supply of dairy products. The price drop has triggered a rethinking on the future shape of global dairy markets. The factors that are expected to re-shape in the dairy sector are :

1. A shift in the global trade away from bulk milk commodities such as SMP, butter, butter oil towards value added products such as cheese, yogurt, deserts and specially milk powders.

   The sophisticated consumer now sees food particularly

dairy products as a means of health and happiness. He is looking for delicious chat are delightfully tasty and healthy as well as authentic and exotic.

2. The developed world is reviewing its policy focus on surplus bulk commodities with a motive provide subsidies for lightening fiscal burden on the developed countries. Incentives may have to be given to develop dairy markets as well as production of value added milk products.
3. The growing competitiveness, triggered by the WTO, has made the global dairy markets increasingly complex. Each year, a large number of new food products are being added in the market place in response to the consumer's search for delicacies that are novel but natural with a touch of mystery and class.
4. Since initiation of economic reform impressive advances have been made in mechanised production of wide range of indigenous milk products. Their large sale production involves intensive engineering inputs for designing the plant layout, selection of various equipment and manufacturing process.

India's traditional dairy products sector is poised for rapid expansion with the result of application of modern process technology in the production of indigenous sweets (Mithais). The rising demand for packaged, fresh products like dahi, paneer, lassi and cream is widening the base of modern dairy sector.

Significantly, technological innovations have come at time when ethic milk products are attracting worldwide attention, as is documented in this studies. This overview focuses on technical and economic development that are transforming this age-old sector.

Market opportunities in India and abroad have also been emphasised. Markets for traditional Indian mineral products also exist overseas where ethnic population of the Indian sub-continent has settled.

However, to tap these markets, it is essential to confirm to the internationals standards of food quality and safety. It calls for Good Manufacturing Practices (GMP), ISO Quality Management System, Hazard Analysis Critical Control Points (HACCP) and food standards formulated by FAO/WHO codex Allimentarius Commission. In step with this trend, the Government of India has made the pre-shipment inspection compulsory for export of dairy products. All these factors eventually lead to the ever increasing emphasis on industrial production and packaging of indigenous milk products.

With the growing consumer awareness towards health and nutrition packaging and nutritional labelling have become important. This trend has been further accelerated by the changing dietary habits and life-style to the ever increasing number of nuclear families. They are demanding convenient, easy-to-cook; ready-to-eat food stuffs in appropriate packaging that return freshness, flavour and taste, preserves nutrition and also a long shelf-life. This is borne out by market increase in expenditure on meals away from home as well as on packaged foods, purchased during regular grocery shopping.

VISION 2020 OF INDIA

Ex-President of India Dr. A.P.J. Abdul Kalam has drafted a *Vision of India* for marketing India as a developed economy by the year 2020 and *Vision Bihar* by 2015 as developed state, which has been called as '*India Vision 2020*'. He has a dream and vision to see India one of the developed countries of the world instead of a developing country as we have every thing required for the same. To him, "What makes a country developed?" The obvious indicators are the wealth of the nation, the prosperity of its people and its standing in the international forum. There are many indictors regarding the wealth of the nation the Gross National Product (GNP), the Gross Domestic Product (GDP), the balance of payments, foreign exchange reserves, rate of economic growth, per capita income, etc. Economic indicators are important but they provide only a part of the picture. He expressed similar view in terms of Bihar while addressing joint session of both legislative houses in March 2006.

Regarding people and development/human development Dr. Kalam viewed, "Many permeates are utilised to indicate how will people are fed; their overall nutritional status, the availability of good nutrition during various phases of their growth and lives, the average life expectancy, the infant mortality rate and availability of sanitation, the availability of drinking water and its quality, the quantum of living space, broad categories of human habits, the indices of various diseases, dysfunctions, disorders or disabilities, the access to medical facilities, toleracies, the availability of schools and educational facilities, various levels of skills to cope with fast changing economic and social demand and so on."

He further stated and referred Gandhiji "one can including many other indicators of the quality of life. Still there is a ragging worry when we apply the talisman prescribed by Gandhiji. Gandhi's strikingly simple criterion was that every action proposed or contemplated should in its implementation wipe the tear of a poor and downtrodden person".

According to Dr. Kalam, "Globalisation, which means integration with world economy, brings the affluence of external forces into our society. . . . We also share the view that competition, both internally and with other global players would be useful to make the country efficient and strong". But unfortunately, developed countries have set-up several non-tariff barriers which strike the roots of ideal. Competition based on market forces. These are mostly aimed at denying opportunities to other countries to reach a developed status.

Even when one country prepares to cope with a set of barriers introduced by these developed countries, either through their own laws or through multilateral treaties, a new set of complex barriers crops up.

"India has to be prepared to face such selectively targeted actions by most powerful players even when it traits to march ahead to realise its vision of reaching a developed status."

"Issues of national security are no longer simple consideration of defence but are closely interwined with many aspects of trade, commerce, investment as well as creation and use of a knowledge-base." It appears to be emerging a new kind of war fare. "If a country does not learn to master these new realities of life, all our aspirations to ensure the prosperity of our people may care to rough."

A developed India should be able to take care of its strategic interests through its internal strengths and its ability to adjust itself to the new realities. For this it will need the strength of its healthy, educated and prosperous people, the strength of itself economy, as well as the strength to protect its strategic interests of the day and in the long-term.

With regard to vision for nation Dr. Kalan Stated, "Nations are built by the imagination and untiring enthusiastic efforts of generations. One generation transfers the fruits of its toil to another which then takes forward the mission. As the coming generation also has its dreams and aspirations for all nations' future, it therefore, adds something from its side to the national vision which the next generation strives hard to achieve. This process goes on and the national climbs steps of glory and gains higher strengths.

The First Vision

According to Dr. Kalam, the first vision is clearly of national vision which constantly drives the people towards the goal. Our last generation, the glorious generation of freedom fighters, led by Mahatma Gandhi and many others set for the nation a vision of free India. This was the first vision, set by the people for the nation. The unified dedicated efforts of the people from every walk of life won freedom for the country.

Dairy food forms an important constituent of the daily menu for people the world over :

- The structure of the dairy in developed countries is characterised by high per capita consumption and slow growth rate.
- Broadly speaking, demographic, income and taste preferences drive the demand for food items like milk production.
- While low fat and skimmed milk are replacing whole milk. This category is facing intense competition from soft drinks and fruits juices.
- Value added variants such as flavoured or vitamin riched milk have not been able to completely counteract the trend.

- Another threat is from non-dairy soya milk.
- Ice-cream continues to be a mass-market dessert in the developed world.
- India has seen double-digit growth in Ice-cream. Thanks to the effort of *Amul,* which made it affordable for the common man.
- According to Sri Harish Damodaran, "Today, only the Gujarat Co-operative Milk Marketing Federation (GCMMF) or Amul is really well-equipped to face the challenges of liberalising and globalisation".
- The setting up of joint venture of National Dairy Development Board (NDDB) with individual state dairy Co-operation Federations through the subsidiary Mother Dairy Food Ltd. (MDFL) would undermine the country's co-operative dairy movement.
- The humble Srikhand of Gujarat has found a fan following in New Zealand. Rasogullas are tingling the taste buds of Americans and gulabjamuns have forced their way during tables of Shikhs in West Asia.
- If international community markets are known to change with kalaidoscopic rapidly, trade in milk and dairy products can be no exception as Indian Milk delicacies make their presence felt overseas.
- It is a miscounception that our dairy products are not doing well in the international market.
- India has the potential to become one of the leading players in the milk and milk products exports since we have the geographical advantage of being located amidst major milk deficit countries in Asia and Africa.
- Major importers of Milk and Milk products are Bangladesh, China, Hong Kong, Malaysia, Japan, Singapore, UAE, Oman, Philippines and other Gulf countries, all located close to India and we should encash by tapping the export potential.
- There is also a vast market for the export of traditional milk products such as ghee, paneer (dairy cheese) and ethnic sweets besides value added milk products such as chocolates, flavoured milk and butter to a large number of Indians scattered all over the world.

- Dairy industry is also the single largest contributor and the country's gross domestic product (GDP) and involves as many as 80 million farming households.
- The social impact of developing the dairy industry/ White Revolution has thus been profound.
- Indian dairy industry has long been the largest and one of the most primitive in the world.
- Growing at a rate of 5% per annum, the dairy industry is expected to triple its production in the next 10 years.
- However, with the opening up of the Indian market to an influx of foreign goods, concern had been expressed over the fate of Indian dairy industry.
- Under the WTO regulations, all developed countries which are among big exporters are supported to withdraw support and subsidy to their domestic milk sector.
- Though WTO framework is based on free trade, European Union and the US have bypassed and even openly violated their WTO commitments.
- Under a scenario where international prices are lower than domestic process, readers can make quick profits without any big investment. But the interests of a large number of formers with critical dependence on dairy may be totally forgotten.
- While developed countries may have evolved better institutional mechanism to protect the interests of farmers and processors, this is not the case with India.
- The free world trade regime ushered in by the WTO poses many threats to Indian Dairy industry but at the same time, it opens up many vistas for opportunities for industry.
- Since the country is no longer facing milk shortage, innovative use of the technology and proper resource management can help the industry achieve export competitiveness in terms of prices yield and quality.
- The Kerala Co-operative Milk Marketing Federation Ltd. (Milma) and NDDB subsidiary, Mother Dairy Food Ltd. have entered into a joint venture (J.V.) to market Milma products in Kerala.

- The J.V. agreement signed in Thiruvananthapuram on December 26, 2002, stipulates that Milma will continue to be responsible for the procurement, processing and packaging of milk and milk products, while the joint venture company will be in change of marketing their products.
- The focus of the new marketing company will be on increasing the availability of Milma milk.
- This will involve boosting the brand's distribution and also improving its cold chain.
- Pricing will be done in such a fashion as to give the consumer the best price, while ensuring that the dairy farmer also gets the best possible price.
- From counting around for suitable dairies to acquire, to intensifying efforts to deliver Paras Products at our door step the Paras group has lined up a slow of activities in the forthcoming year towards increasing its presence in the national dairy scene.
- Touted to be the second largest supplier of poly pack milk in the capital of Delhi, after Mother Dairy—Paras is actively looking towards dairies that it can acquire, particularly in other metros, to establish national presence.
- In India 70% of milk comes from small/marginal/landless farmer in India, cows are more equitably distributed than land.
- Milk has been an important factor is reducing rural income disparity.
- Milk is an important part of the Indian diet. We have been consuming milk and milk products for thousand of years.
- Milk is a unique biological secretion of the mammary gland, endowed by nature with nutrients to fulfil the nutritional needs of the offspring.
- Milk is closely associated with reproduction among mammals and its yield and composition vary among various species to meet the post-vital growth requirements of their offspring.

The Second Vision

The next generation has put India strongly on the path of economic, agricultural and technological development. But India has stored too long in the line of developing nations. Let us, collectively, set the second national vision of Developed India.

Development means the major transformation of our national economy to market one of the largest economies in the world, where the countrymen live well above the poverty line, their education and health is of high standard, natural security reasonably assured, and the core competence in certain major areas gets enhanced significantly so that the production of quality goods, including exports, is rising and thereby bringing all round prosperity for the countrymen.

We have to build and straighten our national infrastructure in an all round manner, in a big way. Therefore, we should build around our existing strengths including the vast pool of talented scientists and technologists and our abundant natural resources. The manpower resources should be optimally utilized to the harness health care, services sector and engineering goods sectors. We should concentrate on development of key areas namely, agricultural production, food processing, materials and

TABLE 8.1
Projected Grains Imports in 2000 and 2010

Countries	*Millennium*	
	2000	*2010*
South Asia	9.2	12.8
East Asia	31.4	39.0
India	6.9	14.1
Pakistan	2.1	4.5
Indonesia	5.7	7.6
China	11.3	21.6

Source : TIFAC, Food and Agriculture Technology Vision, 2020. INDIA 2020 vision for the New Millennium—A.P.J. Abdul Kalam and Y.S. Rajan.

also on the emerging areas like computer software, biotechnologies and so on.

India's needs are very clear, to remove the poverty of our millions as speedily as possible, say before 2010. To provide health for all, to provide good education and skills for all, to provide employment opportunities for all, to a net exporter and to be self-reliant in national security and build up capabilities to sustain and improve on all these in the future.

TABLE 8.2
Projected Household Demand for Food in India at 7% Income Growth

Commodity	*Annual Household Demand (Million Metric Tonnes)*				
	1991	*1995*	*2000*	*2010*	*2020*
Foodgrains	168.3	185.1	208.6	266.4	313.0
Milk	48.8	62.0	83.8	153.1	271.0
Edible Oil	4.3	5.1	6.3	9.4	13.0
Vegetables	56.0	65.7	80.0	117.2	168.0
Fruits	12.5	16.7	22.2	42.9	81.0
Meet, Fish and Eggs	3.4	4.4	6.2	12.7	27.0
Sugar	9.6	10.9	12.2	17.3	22.0

Source : TIFAC Food and Agriculture : Technology Vision 2020. A.P.J. Abdul Kalam and Y.S. Rajan, *India 2020.* A Vision for the New Millennium (1998) Penguin Books Pub., New Delhi.

As obvious from the foregoing table Dr. Kalam has predicted about foodgrains milk, edible oil, vegetables, fruit, meat, fish and eggs and sugar availability in India by 2020. According to him,

(a) A growing population India's population is projected at 1.3 billion by the year 2020.
(b) Economic growth is another factor. As economy group, people earn more and consumption rises. It will be a happy day when our poor have enough to eat.

(c) In addition, there is a definite change in lifestyle, there is a clear trend towards consumption of meat products with the increase in income. Consumption of non-vegetarian food tends to increase consumption of cereals as well.

He has given a scenario of prediction of foodgrains in following Table 8.2.

The growing demand for foodgrains, vegetables, fruits, milk, poultry and meet as well as cash crops is going to present greater and newer challenges to agriculture.

Dr. Kalam and Rajan has suggested following several immediate steps to ensure stability in production. President Kalam has given a brief overview of the vision 2020 in India to envisage India people :

1. India should become a developed nation by 2020.
2. A developed India means that India will be one of the five biggest economic powers, having self-reliance in national security. Above all the national will have a standing in world economic and political fora.
3. To achieve this status, several steps are to be taken in agriculture such as making eastern India a wheat granary and increasing the use of hybrid rice is also for improving the quality and yield of various crops, vegetables and other products. Environmental considerations in agriculture gain attention.
4. Catalyse on the agricultural core strengths to establish a major value adding agro-food industry based on cereals, milk, fruits, and vegetables, to generate domestic health. Also make India a major exporter of value added agro-food products. Agro-food industry and distribution should be absorb a number of persons rendered surplus from increasingly productive and efficient agriculture.
5. A number of engineering industries and service business to grow around the agro-ford sector.
6. India to capitalise on its vast mineral wealth to emerge as a major techno-industrial global power in various

advanced and commercial materials, steel, litanium, aluminium rare earths, etc.

7. Indian chemical industry to be transformed into a global technological innovator in clean processes and speciality chemicals and new drugs and pharmaceuticals, a major business should be created in natural products. Vast biodiversity should be transformed into wealth of people and the national through selective technological interventions. Indian marine resources are to be transformed into economic strength.
8. There is to be a resurgence of Indian engineering industry, machine tools, textiles, foundry, electrical machinery and transport equipment.
   India is to become as not exporter of technology by 2010 as their areas and an important world leader is software for manufacturing and designed also a key contributor to the field of flexible manufacturing and intelligent manufacturing.
9. India should emerge as a global leader in services sector with its vast and skilled human resource base being a come strength.
10. Excellent health services must be provided in India.

Thus, farmers and machines should grow together. India has many such able people, dedicated youths and NGOs.

CULTURED/FERMENTED PRODUCTS

Fermented milk constitutes a vital component of human diet in many regions of the world is also in Indian sub-continent, i.e. dahi (curd) and lassi (squired yogurt), chhachh (butter milk) conservation of milk into dahi/lassi imparts a thicker consistency smooth texture and a distinct flavour that provides safety, portability and novelty to milk nutrients.

Dahi

There are numerous references to dahi in the vedic literature particularly in 'Rig Veda'. In India, dahi is largely made and eaten at home using traditional kitchen recipes. Dahi is generally consumed in its original form to the meal and may be consumed as a sweet or surveys lassi drink or as a dessert containing sugar and fresh diced banana orange slices, mango fits and other seasonal fruits.

In general, dahi contains none protein, calcium and other nutrients than milk. Dahi, Lassi and cultured milk are made from fat standardized milk. All ingredient should be of high bacteriological quality. Its technology, equipment storage tank, heat treatment, homogenization, fermentation, pacuring and storage and quality control need further and done monitoring.

Misti Dahi

For centuries, Misti dahi also known as payodhi and lal dahi, has been a well established desert in the eastern parts of India. It is a popular dessert in West Bengal. It is a delicacy of choice during religious fricatives. In combination with rasogolla, misti dahi is regarded an extra species desert in rural and urban Bengal. Misti dahi preparation is mainly confined to domestic or cottage scale operation. Misti dahi is a fermented milk product. It is generally prepared from Cow mixed milk. It is coloured and flavoured commonly with caranel. Care sugar is used as most common sweetened. Misti dahi cups should be properly created or cartoned. It is many ways similar in the western dairy product.

Shrikhand

Shrikhand is a semi-solid, sweetish sour fermented milk product prepared from dahi. Like dahi, it is very refreshing, particularly during summer months. Shrikhand is obtained from chakka or skimmed milk chakka to which milk fat is added. It is delicious and delightful dessert of eastern India. It has the nutritive goodness of fermented milk products. It is the first traditional milk products for which large scale production technology was worked out. It has been produced on an industrial scale at sugar dairy is Vadodara, Gujarat since early 1980.

Chakka Powder

Manufacture of Chakka powder involves preparation of dahi followed by partial removal of whey from it. The chakka is then ground is a colloid milk to obtain a smooth and uniform consistency in starry.

Shrikhand Wadi

Shrikhand Wadi, a popular product has its origin in the western region of India. It is an extension of Shrikhand. It is obtained by desiccating shrikhand to a hard mass by heating in an open pan over direct low fire.

Lassi

Lassi (stirred dahi) is a refreshing summer beverage, popular in North India. It is a white to a seamy, white, various liquid with sweetish, rich aroma and mild to high acid taste. It is flavoured either with salt or sugar and other condiments, depending on regional preferences. Lassi is obtained from pasteurized milk or part skim milk. Lassi making was earlier confined to the cottage sector at homes. It was mainly a rural product. Now it is commercially prepared in several parts of North India. Commercial Lassi is various, cultured fluid milk. It is packaged in traditional milk cartoons/boxes. Salted lassi is marketed in a number of citizens in the southern region of the country. The process used in the manufacture of lassi include pasteurization, homogenization and culturing system. Traditionally Lassi is prepared by sterling dahi with a small quantity of cold water. It is flavoured with rare water or kewre flavour and filled in bottles or pouches before refrigerated storage.

Matha/Chach/Chhas

Matha, Chach or Chhas are synonyms for a product buttermilk, which is commonly consumed in all parts of India. It is a popular refresh summer beverage. It is a by-product obtained in the preparation of makkhan from dahi by traditional process which involves changing a dahi/cold water mixture. It is called Moru in Kerala and is consumed as an accompaniment to rice and pickled. Matha is highly flavour and milady to high acidic. It contains 6-7% milk solids and 1-2% fat.

Kadhi

Kadhi is a prominent culinary food item prepared from dahi throughout India. It is made from four sour dahi with addition of wood apple fruit and Indian sorrel lives followed by seasoning with pepper. The composition of Kadhi varies from region to region depending upon consumers preference.

In the traditional method Kadhi preparation 5-8% of Bengal gram Besan is added to stirred dahi or buttermilk. Besan Acts as thinking agent.

Raita

Raita made from dahi and serves as an accompaniment to meals. It is used with rise dishes may also form part of meals where wheat bread (Roti) is the dominant component. To prepared raita dahi is lightly beaten, spices and settled to last. It is prepared fresh at the time of consumption.

Dahi Vada

It is eaten as a snack as a side dish. It is a variation of raita. In its preparation deep fried dal, butter patties are dipped in dahi in stead of vegetable or fruits pieces. Dahi Vada uses Urad Dal or some time gram Dal for preparing Vada. The sald is soaked in over night the drain and ground to thick butter and mixed with spices and shaped into pices. It is covered after fridge in Ghee or Oil completely with dahi with garam mashala, Chilli Powder, Sweet surd, Imali, etc.

Fat-Rich Products

In the traditional Indian dietary regime, milk fat is the form of malai (Cream), Makhan and Ghee physical contributes significantly towards nourishment of people of most all the age groups. The ancient Vedic literature refer to Ghee derived from cow milk as having excellent nutritional and tonic qualities. In Sanskrit Ghee (Ghrit) is described as the food fit for Gods and accomunidity enormous value. Buttermilk is an important by product obtained during conversion of milk into fat rich dairy product.

Ghee

Ghee is popular in Indian continent. Particularly Deshi Ghee is widely used milk product throughout India as the supreme cooking medium. Almost entire range of sweet are prepared by Ghee, Milk, Cereal, Fruit, Vegetable, etc. Ghee production is an important activity, ghee organised socket of Indian Dairy, it is accidental to the dairy product mixed. Ghee is the clarified butter fat obtained cow or buffalo milk. The Gee prepared from Makhan contains many aromatic compounds from during the fermentation of Dahi. The main Madies of Ghee are located at Hatras, Khurja, in. U.P., Porbandar in Gujarat, Guntur in Andhra Pradesh, Erode in Tamilnadu and Jaipur in Rajasthan. It is regularly exported to Bangla Desh, Bhutan, Nepal, Middle East Country and North America.

Ghee Residue

Ghee Residue is useful product of Ghee Industry particularly from Makhan, Cream, etc. It is produced at about 65,000 tonnes per annum enlarge Ghee manufacturing Plant its average yield is about 12% by direct P.M. method, 3.7% by Creamery butter and deshi butter under domestic condition in small skill condition.

Makhan

It is always patronised for it delectable favourable and rich attributes. It is favour as a side dish with Paratha or Make-ke-Roti. It is the traditional unsalted butter, the product obtained is of soft consistency. It yield varied between 4-7%.

Malai (Cream)

It is a traditional symbol of richness. Indian food are combined with it. Rasmalai, Malai Kopta, Malai Tikka, etc. It is readily available at home and is consider superior to chemical best cream. Traditional Malai is also added to hoi thicken suitaned milk as an after milk or on festive occasion. It is produce with different fat content, it may be stored for 1 to 7 days under ambient conditions and 3-2 for weak under referenced condition. It is used in the preparation of ghee in traditional manner

MILK-BASED PUDDINGS

Milk sweets have been in unrepeatable part of the service cultural life in Indian sub-continent.

1. Kheer

Kheer the favourite event per excellence of India, is a sweet milk rice confection. Kheer has the status of royal treat. No feast is considered complete without Kheer. Hindu mythology refers kheer as the celestial nectar, Amrit. The Hindi Word Kheer is derived from Sanksrit, Ksheer. It is popular all over India. As a per-evident milk delicacy it has been associated with festivals and celebrations from time immemorial. Generally, Kheer constitute a homogenous mass of milk and rice as both flow together. Kheer is prepared upon all ritual and festival occasions as an auspicious item of food placate the gods. Kheer-puri is common mans' idea of great feast. It is must in Diwali. It is popular throughout India. NDRI, Karnal has developed its latest technique.

2. Dread Kheer Mix

Production of Kheer is suitable form of reconstitution which food health problems of its self-life. It is intent kheer mixed the milk, rice, sugar, slurry representing the simulated liquid phase of kheer. This produce found to have a high acceptance which has great commercial potential of dairy product.

Payasam

Payasam is mean milk. It is a kind of kheer most popular in South Indian States. It resumes kheer of North India which is prepared of rice and thicken milk occupied by fruits. It has several variety like Khus-khus in Karnataka, Palada Payasam in Kerala and Tamilnadu Shirkurma and Gil-e-Firdaus in Hydrabad cities. It is given with honey and sugar infants as first Unprayasan.

Phirni

It is a delicious variety of Kheer prepared by cooking milk with rice paste and sugar. It generally covered with silver skim-milk in Hydrabad. It is prepared with pineapple bits.

Sevian

Sevian is a sweet dish in North and South India with both sabudana and sago granules are used to make food for invalids like Phalluda and milk best payasam. It is consumed traditional festivals like Navratri Ikadashi, etc. It is commercially valuable.

Lauki Kheer

North India Lauki Kheer is one of the most popular vegetable based milk delicacies. It is generally produce to domestic level. It is good source of carbohydrate, Vitamins A and C and Minerals in Kashmir in Lauki Kheer, Rice, Flour and custard powder are also added.

Sohan Halwa

Its production confined to North India. It product profile differ from region to region and from Halwai to Halwai. It is a wheat-based product with milk, sugar, Ghee and Khoa.

Gazar Ka Halwa

It is very popular in India particularly in winter season Gazar is valued for its growth Protein, Nutrients and is produce household level and by local sweet maker on small and cottage scale in form of sweet Halwa, Barfi, etc. which are most popular in the market.

Kazu Barfi

Due to delicate texture good flavour and excellent mouthful it is most popular in whole India. It is always in demand which is exported in middle-east and European countries where it tend important its halwa and sweets are very popular.

Heat-Acid Coagulated Products

Channa and Panner are two prominent traditional coagulated milk products of Indian sub-continent. Both provide sound nutritional variety, convenience, prominent and softy as well as novelty and texture to consumers. Its main products, in addition to panner and Chana are Rasogulla, Rasomalai, Rajbhog, Kheer Mohan, Sandesh, Chhana Murkhi, Chamcham, Channa Podo, Surti Paneer, Bandel Cheese, regional products like Kheer Sagar, Sitabhog, Chhana Gaza, Chhana Jhale, Chhana, Kheer, Chhana Pakoda, Rasbali, Shoscim and Kalari, etc.

Desiccated Milk Based Product

It product in based of Khoa which occupy in prominent place in the preparation in dairy market like Barfi, Peda, Gulabjamun, Kalajamun, Pantua, Lalmohan, Kalakand, Milkcake, Bal Mithai, Dharwar Peda, Rabri, Khurchun, Basundi, Kulfi, Falludda, etc. These Khoa-based dairy products are rich source of energy.

COMPFED

Everything in touches, terms into gold. The 'Midas touch' story is one of the Bihar State Co-operative Milk Producers' Federation Ltd., (COMPFED) which has been scaling new height of screen in recent times. Be it milk, curd, panner or spongy rasogulla or ice-cream, peda, all products the COMPFED have created a brand for themselves in the Bihar market. No wonder, the Government of Bihar has been showcasting the success story of the COMPFED in a big way to outside world." According to MD of COMPFED Mr. Girish Shankar, "Our turnover figure exceeded beyond Rs. 400 crore during the last fiscal year 2004-05. But this time around, we expect to clock an annual turnover market of nearly Rs. 450 crore".

Expressing confidence Mr. Girish Shankar has stated that COMPFED was poised to take a huge keep in the coming years. Recently, vegetable grower cooperative was also being proposed. He further said COMPFED, which sells its products under the brand name Sudha, will soon come out with its own pack of honey. The Sudha honey will be far superior in quality and more importantly it will come at a cheaper price than what one was paying for other brands of honey."

Truly, COMPFED has launched itself into an expansion mode. Even today, it cames out with its new launch the Sudha brand cow milk and Ramdana Lai a popular ethnic sweet which will be another addition in Sudha products. "Similarly there is high demand for cow milk".

So far as Bihar is concerned the contribution of dairy sector has been increasing during last 3-4 Plans particularly during VIII, IX, Xth Plans in the state; despite less government expenditure and very poor priority given by Bihar Government to the sector.

It has been observed from this study that COMPFED is playing very prominent role in fulfilling its objectives since 1983 in most backward state like Bihar.

(a) COMPFED is carrying out the activities to prompt production, procurement, processing and marketing of milk as well as dairy product.

(b) It is trying to develop and expand other allied activities as may be conducive to promotion of Dairy Sector.

(c) It is actively engaged in purchasing commodities from member and from other sources also.

(d) It is trying to study problems of mutual interest related to dairy production procurement, marketing and processing of dairy product.

(e) Regularly supply of milk production has been instrumental in meaning the confidence of farmers and improving productivity of animals.

(f) Today Sudha household name in milk and milk products. It is first country to introduce sugar free ice-cream aimed a diabetics and health conscious people.

(g) COMPFED has been manufacturing various category of Milk, i.e. full cream milk (Sudha Gold), Standard Milk (Sudha Shakti), Tonned Milk (Sudha Healthy), Double Tonned Milk (Sudha Smart), Skim Milk (Sudha Light).

(h) COMPFED has 7 milk sheds in Bihar and 3 in Jharkhand namely, Baruani, Mithila, Tirhut, Sahabad, Vishal Patliputra, Gaya, Bhagalpur in Bihar are in under direct control of COMPFED.

(i) The coverage districts of Milk shade of Bihar are Begusari, Khagaria, Lakhisarai, Shekhpura and some villages of Munager and Saharsa.

   (i) Mithila with the headquarters at Samastipur are: Samastipur, Darbhanga and Madhubani districts.

   (ii) Shahabad milk-shed with the headquarters at Arrah are—Bhojpur, Buxar, Rohtas and Kaimur.

   (iii) Tirhut milk-shed with the headquarters at Muzaffarpur are Sitamarhi, Deghar, East & West Champaran districts.

   (iv) Vishal Patliputra milk-shed are rating Vaishali, Nalanda, Siwan and Siwan district.

(j) There were as many as 4018 Dairy Co-operative Societies (DCSs) at the end of March 2003, of which 3285 were functional while 1647 more registered. Of the 3285 functional DCSs as many as 2598 are covered under SIS and COMPFED performed 5.36 lakh inmentation during 2002-03.

(k) Capacity building/skill upgradation through framing and management have been given maximum emphasis in implementation of dairy development programmes in Bihar. Consequently, now COMPFED had entered fruit and vegetables business and exported titch.

Besides, COMPFED has been providing self-employment to about 3 lakh farmers and has been able to make an input on the prevailing societies system. Producer members belong to all castes and creads come together to a common platform DCS to pour milk and all of have equal valuing night.

(i) Further, COMPFED and unknowns have implanted other of socio-economic programmes, namely, rural family health project with the support from certze for development and population activities of USA. With support from Bihar Rural Energy Development Agency, Programme of Work Empowerment with support from the Development of child, the ministry of Human Resource Management and programme of poverty evaluation with support from the finance and development co-operation for SC/ST and OBCs

(ii) Excellent work have been carried out by the COMPFED in Bihar which have been recognised by the various agencies and received following awards, i.e. National Award for Dairy Development and production, Best Productivity award from National Productivity, New Delhi. In addition to providing ready and stable market for rurally produced milk, union has been providing milk production enhancement inputs like artificial insemination with frozen seman, Veterinary First-Aid, Vaccination, supply of balance feed, Folder seeds. Paddy and wheat Bhusa with urea.

In tune with change with general environment, COMPFED has respondent favourably for producing quality milk and milk products. Dairy Plant Management Programme was launched in

1992 has followed by quality assurance programme (QAP) in 1993. This has resulted in bringing about a positive change leding to viability coupled with decreasing operational cost and improved quality of products. Future area of task head of COMPFED are making sudha brand a global market leader, consolations of DCSs, leading to increase procement, improvement in active involvement and participation of member producers as well as sell of both milk and milk product and optimum utilization of plant capacities.

COMPFED has five Milk Unions with its direct control, covering 23 districts in Bihar the total installed capacity of COMPFED 7.5 Lakh letter/day. It has over 4500 DCSs with a number of 2.50 lakh. At the aggregate 56% of installed capacity are utilised. The products of COMPFED under the brand name Sudha as also recognized importance of quality and 9 out of 10 varies have all redivide accredit with ISO 9001:2000 and HACCP ISO 1500 : 1998. Of the total DCSs of COMPFED Vaishali Patliputra has the highest number. Barauni has the highest procement. As a whole, financial heath of COMPFED is satisfactory.

As such, in a depression where almost all states promoted corporation are helplessly swimmingly in the Ocean of Red, COMPFED is the only glittering star on the industrial horizon of Bihar. Even in north-east Indian regions, at present, COMPFED is handing about 10% of marketable surplus living scope for organization, the programme is uncovered villages of Bihar.

Thus, providing gainful employment to women as always a concern of policy-makers in Bihar. Experiences of Anand pattern came hardly and dairy movement involving women are identified as a major plank. Consequently, Bihar Women Dairy Project launch was 1998 to provided gainful employment and to increasing village community. During 1998 to 2005, Bihar has the privilege of implementing four phases of WDP during 1998 to 1990 in first phase the union covered by Barauni, Mithila, Patna and Tirhut. During 1991-93 Second Phase Baruani, Mithila, Patna and Tirhut Union. During 1996-97 in third Phase Barauni, Mithila, Panta and during 1998-2005 in fourth phase during 1998-2005. Barauni, Mithila, Patna, Ranchi, Shahabad and Tirhut Union were covered.

This study addresses the position of Dairy Sector in WTO under:

(a) Agreement on Agriculture.
(b) Agreement on Trade Related Aspects of Intellectual Property Rights.
(c) Agreement on Application of Sanitary and Phytosanitary measures.
(d) Agreement on Technical Barriers to Trade.
(e) International Dairy Agreement.

It addresses three pillars of *Uruguay Round of AoA (URAA)* i.e.—*(i) Market Access, (ii) Domestic Support, and (iii) Export subsidy*. It suggests policy options for domestic market of the reforms and next round of WTO negotiation. The issue of global competitiveness has been addressing through policy analysis matric (PAM) by calculating domestic resource Cost Ratio DRC under various Ministerial Conferences of WTO particularly in Doha (2001), Cancum (2003) and Hong Kong in 2005 in NDA and UPA Governments.

It has also been analysed the traditional major of competitiveness namely, nominal production co-efficient, effect product and effective co-efficient. It has also been assumed the domestic support in the forms of three controversial boxes—1. *Amber Box,* 2. *Blue Box,* and 3. *Green Box* subsidy being provided to dairy sector all over word.

Therefore, India along with other developing nations needs special safeguard against distorted world prices. Export enhancing measures by developed nation shall we forbidden. Farmers in developing countries need rules that outlay the expert of agricultural goods. As the prices below received by the producers. Those rules must extent beyond direct subsidy to cover the full range of major. Blue Box major should be limited real decomplying of payment and production is needed. Rich country need to open their own markets.

India must have greater participation negotiating table and must have their own exploit agenda about tariff reduction of domestic products by major developed country. Further, we should intent for reduction in production related export credit and development transpiration discipline for export credits. The specific actions will help in establishing a more equitable system of global trends of dairy products.

Over 50% milk produced in India consists of Buffalo milk. Buffalo milk gives greater out tern than cow milk because it has higher fat and total solid content, proteins, minerals, etc. Despite being largest producer in the world India has little experience and is a miner player in the world market. The total Indian consumption accepted increasing 60 million tonnes in 1997 to 140 million tonnes in 2020. As per capita consumption is accepted to rice from 61 kg. to 104 kg. during the same period.

- Under new WTO regime, as dairy sector moves towards less government is intervention and regulation, doubts are raised about competitiveness and efficiency of our dairy industry by quarter corners with the future directions of the dairy industry, trade in a liberalised and global framework.
- However, very little efforts have been made so far to find out the performance of our dairy industry in terms of its capability to face comparations in the open and liberal framework.
- Under WTO regime particularly at Ministerial Conference held at Hong Kong in December 2005 the following issues need special attention :

(a) How our dairy industry has performed terms of production, consumption and trade during last 56-57 years?

(b) What are the future prospects and challenges before our dairy sector?

(c) What are the main provisions and their impacts of liberalising and globalising world trade in dairy products?

(d) Whether it has been successfully in reducing barriers to trade and market distortions and what are the problems that have emerged during a implementation of Agreement on Agriculture (AoA).

(e) In India dairy industry globally competitiveness in an open economy environment. What are the impacts of various factors that influence competitiveness of dairy sector?

Suggestions

On the basis of above analysis and observation we may suggest the following :

- India must make concerned effort in synchronization of Indian milk production system with OIE requirements based on strategies.
- The global economy today centres around technology transitional production system and international competitiveness.
- Standardization and quality management system play a major role in technology effecting economy in production and competitiveness.
- The emergence of global market of unprecedented magnitude has created a demand for globally standardized dairy products.
- The 21st century belongs to consumer. Internet revaluation, e-commerce, and abolition of trade barriers through WTO, has reduced the world to a global village in the true sense. The consumer has plenty of options now. Therefore only the bent is terms of quality, service and cost is going to last the competition.
- Consumer must be contracted for their comment about the milk product quality to locate deficiencies if any for further improvement. All cases must be attended with proper care.
- The Technical requirement during dairy product distribution network must be considered by the organization for safe marketing and be included in the quality manual as operational guide.
- Under this changed situation, positive scientific distribution management system is required to the exercise for safe distribution, stocking and sale of dairy products. Accordingly, adequate facilities must be ensured at every level of marketing network.

- In order to improve quality of milk constituents different DNA-based technologies are available. Among these, marks assisted section (MAS) is quite promising and has an edge over others, due to its capacity to manipulate animals genetic make-up for production/reproduction disease resistance along with milk quality traits in diary animals.
- Because of its commercial and nutritional significance, milk has been considered as an ideal known food with respite to its synthesis, composition and structure of individual components.
- Milk contributes 3.1% protein, 4.0% fat, 5.0% lactose, 0.74% minerals and a sizeable amount of vitamins etc.
- The most variable component of milk is protein. Milk protein composition differs qualitatively and quantitatively among different milch animals spices.
- In order to improve quality of milk constituents different DNA-based technologies are available.
- Considering the wide variations in the socio-economic and geoclimatic conditions under which the dairy animals are reared, vary differently in different parts of the world.
- With the beginning of the new millennium, the role of marketing in the milk business assumes greater significance.
- Marketing liquid milk is different due to a number of reasons, namely, product, characteristics, variants, usage patterns, consumer perception and behaviour, distribution practice and retailer's profile.
- Milk a primary commodity, is also highly perishable. It has limited self-life even after processing. Consumers have strong preferences with regard to milk in terms of flavour, etc.

  In terms of usage, milk is an item of daily consumption and perceived to be essential.
- Distribution of Milk requires a dedicated setup *vis-a-vis* distribution, procession and sales force.

Govt. of Bihar and India must try to simplify the procedure for recessing funds. They should also insist on conduction of concurrent evaluation while the projects are in progress.

Under no circumstance Milk Unions in Bihar should not be allowed to organise societies after fifth year of a particular phase. It should also direct federations to conduct regular review committee meetings, both at state and district levels and proceeding should reach in a fortnight. Further, Government should make effort to organise all India conference for officials of the dairy federation and the milk unions involved.

Milk unions should strengthen animal health support by presetting private veterinary doctors on a pilot basis. Unions may like to provide greater emphasis on group developments by involving local NGOs.

Secretaries need to be trained in promoting under dimensions of woman's societies and economic development. They should provide greater emphasis on gender sensitisation in training of project staff.

The link agencies of dairy sector like department of education, health, animal husbandry, industries, NGOs. banks, etc. should consider work allocated as part of their activity and honour it. They should also attend the review meetings and camps/training progress as agreed upon. Linkage agencies must extend whole hearted co-operation to women and unions so as to make the exercise a grand success.

Economic motivation, extension, contact, commitment towards dairy co-operatives, distance from chilling centre and milk plant, quantity of cattle feed supply, daily milk collection business, milk price, performance of village milk produces co-operative socialites (VMPCSs) etc. the main factors for the success of VMPCSs must be carefully taken into account.

In order to economies the line, efforts and money organisations of VMPCSs must be judged against agricultural resources and village infrastructure at the command of a village.

High level managers and concerned staff must work honestly and try to reduce the main constraints of dailing and co-operative and improve the performance.

The main constraints like lack of efforts towards milk production by VMPCSs and DCMPUs (District Co-operatives of Milk Produces Unions). Low economic motivation of members-farmers, and artisans in village and other related areas, scarcity of water and greenfodder, low risk bearing stability of dairy farmers and delayed veterinary and by DCMUS must be removed. Every effort must be made to remove the serious constraints perceived by the employees of milk co-operatives, differences in policies at the DCMPU and VMPCSs, levels, seasonal variation of process of milk products.

In order to remove these bottleness, batter co-ordination, decentralised decision-making, provision of rewards on the basis of performance and flexibility in the working of milk plants must be incorporated in the bye-laws. In order to make COMPFED the viable and efficient and to complete, the Milk Unions should have separate cell for monitoring progress of milk co-operatives, milk product and *second wave of white revolution.*

Present WDP and COMPFED cell at the federation level should be strengthened in Bihar with adequate staff and they should be allowed to continue till a phase is completed. Federation, in its part, should release grants on regular basis and insist utilisation and progress reports from the respective unions. The view of milk unions should be taken into consideration while formulating or modifying project proposals.

Federation should also advice milk unions to appoint lady staff as resource persons on productivity linked basis and to absorb then soon after completion of project wherever possible. Federation should also keep a close supervision on organisation and performance of WDCSs in the milk unions. It should also submit progress reports and fund utilization certificate on regular basis in the prescribe formats. Milk federation must also organised state level review committee meetings regularly and follow up decisions taken in these meetings. Emphasis should be on organising Self Help Group (SHG) in dairy co-operative and COMPED. In Bihar, steps must be taken to garner funds from various cooperation and groups for strengthening programme.

Dairy farming in Bihar must introduce an opportunity to interact directly with officials of the milk unions, veterinarian and officials from departments of health, adult education, industries, and NGOs.

For an effective delivery of results all sub-systems should work effectively to achieve overall objectives.

Farmers should be educated on various aspects of dairy farming complete. with creation of proper infrastructure facilities which should help in increasing the adoption of technologies. Care need to be taken in identification of suitable farming system location while introducing the technologies.

There is a greater need of formulating Central Dairy Herd Association and Bihar State Dairy Herd Association with regional branches consisting of consultants, experts, stockness, testers who should make a periodical check-up on all dairy farms for quality and quantity of milk. These reports must be examined by the experts. The finances of such Association must be borne by Government and partly through the compulsory membership on the basis of animals on each forms.

The dairying must be helpful in removing poverty and unemployment and also in checking unwanted migration of rural poor. Milk yield is largely dependent upon how milking is done. To ensure effective milking all excitements or disturbances to the cow should be avoided. Stimulation should be provided for full left down of the milk and milking should be done gently but quickly.

Measure should be taken to protect the cow from flies and practices. Her body should be clean. The cow should be allowed to move as freely as possible and given fresh and clean water to drink. The utensils should be sterilized by heat and chemicals.

The dairy industry needs a stable milk fat to facilitate storage and transportation and to even out seasonal variations in milk production. The butter-oil is more convenient to handle them better, as it is more homogeneous and easier to proportion.

The cost benefit analysis in dairying must be taken in to consideration alongwith all the dimensions of the programmes for a complete evaluation of an extension programme.

The approach of the Indian dairy sector must be to provide more employment and income opportunities to the marginal farmers, small farmers, agricultural labourers and also the educated unemployed through dairying in Bihar and all the States. This subsidiary occupation must play a vital role in improving the rural economy, absorbing large number of agricultural labourers and rural people, abolish the disguised unemployment, eradicate poverty and inequality of income distribution.

Co-ordinated multi-disciplinary research projects must be undertaken with the object of increasing the productivity and the product of dairy sector in the days to come particularly by 2015. Dairy marketing must be properly organised for making dairy products fully remunerative. The major beneficiary like the traditional indigenous middlemen must be removed.

For this there is an urgent need for organised dairy marketing from stage of production to marketing and profitable utilisation of dairy products. Consequently, modern management and technical process in relation to production, processing, transportation, promotion and distribution with a view to improving efficiency at all these levels at minimum cost is quite essential.

The bottlenecks like payment of unfair prices, inadequate improper fittings, limited transport network facility from remove area and poor sanitation in trade of dairy milk and milk product must be remove at the earliest.

The supply system of dairy product should be improved through training and extension, technical inputs, price incentive, existing milk collection route, improving quality and communication systems etc.

We must compete and co-operate in ensuring quality and we should try to continue to educated the consumers about the importance and value of the quality. Efforts should be made to raise productivity and ensure quality and a government policy in this direction is the need of the hour.

The time bond action-oriented implementation policies are required quality enhancement by involving NGOs, Co-operative and private sector. Besides policy and programmes should be based on the actual position of economy of Bihar as highlighted an demonstrate in the white paper issued in the month of March 2006 by present Bihar Government and the measures suggested by the President of India Dr. A.P.J. Abdul Kalam in his historic address in the joint session of Bihar legislative Assembly and council in March 2006 to give new boost to the diary farming in 21st century and making second wave of white resolution a grand success not only in Bihar but in the whole nation.

APPENDICES

TABLE 1

List of 300 Traditional Dairy Products Grouped in Four Major Categories, Produced in 100 Developing Countries, Classified under Four Geographical Regions

CHEESE

Africa

Aoules (Algeria)
Ayib (Ethiopia)
Braided Cheeses (Sudan)
Country Cheese (Nigeria)
Fromage (Madagascar)
Fromage Blanc (Madagascar)
Gibbna (Sudan)
Mashanza (Zaire)
Mboreki Ya Iria (Kenya)
Mudfara Cheese (Sudan)
Pont Belile (Chad)
Takammart (Algeria)
Tchoukou (Niger)
Wagashi (West Africa)
Wagassirou Gassigue Woagachi (Benin)
White Cheese (Benin)

Asia

Chhana (Bangladesh, India, Nepal)
Chhuga (Nepal)
Chhurpi (Nepal)
Churtsi (Bhutan)
Durukhowa (Bhutan)
Kesong Puti (Philippines)
Kimish Panier (Afghanistan)
Paneer (India)
Panir (Pakistan)
Ponir (Bangladesh)
Serkam, Sher, Shergum, Dartsi (Bhutan, Nepal)
Shosim (Nepal)
Tahu Susu Atau Dadih (Indonesia)

Latin America

Altiplano (Bolivia)
Chihuahua (Mexico)
Colonia (Uruguay)
Cotija Cheese (Mexoco)
De Mano (Venezuela)
Farm Chanco (Chile)
Farm Goat Cheese (Chile)
Goya (Argentina, Uruguay)
Guayanes (Venezuela)
Oaxaca (Mexico)
Paraguay (Paraguay)
Patagras (Cuba)
Queijo De Coalho (Brazil)
Queijo De Manteiga (Brazil)
Queijo Minas (Brazil)
Quesillo (Bolivia, Chile, Ecuador)
Quesilto De Honduras (Honduras)
Queso Anidon (Peru, Ecuador)
Queso Beniano (Bolivia)
Queso Blanco (Central America)
Queso Chaqueno (Bolivia)
Queso De Freir (Dominican Republic)
Queso Fresco (Bolivia, Ecuador)
Queso Prato (Brazil)
Requeson (Peru)
Tafi Cheese (Argentina)
Yamandu (Uruguay)

Near East

Akawieh (Lebanon)
Anari (Cyprus)
Awshari (Iraq)
Baladi (Lebanon)
Beyaz Payneri (Turkey)
Chelal (Lebanon)
Daani (Egypt)
Feta (Tunisia)
Fresh Cheese (Lebanon)
Gibbneh Beda (Egypt)
Graviera (Cyprus)
Hallon (Iraq)
Halloumi (Cyprus)
Helloum (Lebanon)
Jben (Morocco)
Kashkaval (Cyprus, Lebanon, Tunisia)
Karichee (Lebanon)
Karish (Egypt)
Kasar Peyneri (Turkey)
Kefalotyri (Cyprus)
Mesanarah (Syria)
Mihalic Peyneri (Turkey)
Mish Cheese (Egypt)
Ma'aimeh (Lebanon)
Raphitoco Cheese (Cyprus)
Rahssr (Egypt)
Shankalish (Jordan, Syria)
Soft Cheese (Iraq, Jordan)
Tulum Peyniri (Turkey)
Umbris (Lebanon)
White Cheese (Qatar, Turkey)

ACIDIFIED MILKS

Africa

Akile Nukadwark (Kenya)
Amacunda (Zaire)
Amasi (Zimbabwe)
Ambere (Kenya)
Arrera (Ethiopia)
Chambiko (Malawi)
Fadhi (Somalia)
Garoor (Somalia)
Hard Femented Milk Curd (Ethiopia)
Ikuvugoto (Zaire)
Irgo (Ethiopia)
Iria Imata (Kenya)
Kadam (Mali)
Kamabele/Kamabou (Kenya)
Lait Caille (Mauritania)
Leite Dormido-Leite Coalhado (Cape Verda Islands)
Mabisi (Zambia)
Mabobo (Madagascar)
Madila (Botswana)
Mariwa (Kenya)
Mashoronga (Zimbabwe)
Mase (Mozambique)
Mazia Maivu (Kenya)
Maziwa Magando (Tanzania)
Mursik (Kenya)
Non Mai Yami (Niger)
Nono Koumou (Burkina Faso)
Nyaame (Ghana)
Pindidaam (Cameroon)
Raib (Chad)
Roab (Sudan)
Rouaba (Chad)
Sawa (Zambia)
Sour Milk (Kenya)
Suusac (Somalia)
Umlaza/Mutuvi (Zimbawe)
Yaourt (Madagascar)
Yoghourt (Nigeria)
Youghurt (Sudan)
Zabadi (Sudan)

Asia

Airag (Mongolia)
Bogurar Dohi (Bangaldesh)
Chakah (Afghanistan)
Dahi (Indian sub-continent)
Lassi (India, Pakistan)
Natural Yoghurt (Fiji)
Susu Madu Klenceng (Indonesia)

Latin America

Boruga (Dominnican Republic)
Dahee (Guyana)
Kefir (Ecuador)
Sour Milk (Nacaragua)
Yoghurt (Latin America)

Near East

Dough (Iran)
Eqet (Jordan)
Jemid (Jordan)
Khather/Robe (Qatar)
Laban (Lebanon, Syria)
Labaneh (Lebanon)
Labna (Iraq)
Labneh (Egypt, Jordan, Libya, Syria)
Leben (Morocco)
Raibb (Tunisia)
Rayeb (Jordan)
Yoghurt (Cyprus, Saudi Arabia, Turkey)
Zabady (Egypt).

BUTTER AND MILK FAT PRODUCTS

BUTTER

Africa

Beurre (Madagascar)
Butter (Nigeria)
Kibe (Ethiopia)
Libonga (Kenya)
Maguta (Kenya)
Mateka (Zaire)
Mauta Ma Med (Kenya)
Mauta Ma Ng'ombe (Kenya)
Mbata (Keynya)
Nebam (Mali)
Siagi (Tanzania)
Sihin/Subag (Somalia)
Sour Cream Butter (Ghana)
Thiagi (Kenya)
Wgi (Benin)
Zubd (Sudan)

Asia

Ma (Bhutan)
Makkhan (India, Pakistan)
Nauni Ghiu (Nepal)

Latin America

Mantequilla (Bolivia, Ecuador, Nicaragua, Peru)

Near East

Smen (Morocco, Tunisia)
Zebdah (Egypt)
Zibd/Zibadah (Qatar)
Zobdeh (Syria)

CLARIFIED BUTTER

Africa

Beurre Traditionnel (Mauritania)
Dine Bagar (Chad)
Ghee (Nigeria)
Mai (Ghana)
Namboumgoun/Nan-an-goum (Cameroon)
Nebammnai (Cameroon)
Sainilli (Kenya)
Samli (Tanzania)
Samin/Semin (Sudan)
Smen (Morocco, Tunisia)

Asia

Desi Ghee (Pakistan)
Ghee (Bangladesh, India, Nepal)
Htaw But Si (Myanmar)
Minyak Samin (Indonesia)

Latin America

Ghee (Brazil, Guyana, Trinidad)
Manteiga De Garrafa (Brazil)

Near East

Maslee (Iraq)
Roghan (Iran)
Samn/Samnah (Egypt)
Samneh (Jordan, Lebanon, Qatar, Syria)

OTHER FAT PRODUCTS

Gaimar (Iraq)
Gibde (Chad)
Keshda Mosakhana (Egypt)
Shmen/Semna (Algeria, Mali, Niger)

OTHER MILK-BASED PRODUCTS

Africa

Amavuta (Zaire)
Chak Mapuo (Kenya)
Mkango (Kenya)
Nyuka Mar Chak (Kenya)
Omokora (Kenya)
Silmissafande-Katare (Burkina Faso)
Umthubi/Munhamba (Zimbabwe)

Asia

Basundi (India)
Dodol Susu (Indonesia)
Ezgiy (Mongolia)
Gundpak (Nepal)
Kembang Gula Susu (Indonesia)
Khawa (Nepal)
Kheer (India)
Khurchan (India)
Krupuk Susu (Indonesia)
Kulfi (India)
Kurchan (Philippines)
Leche Flan (Phillipines)
Malai (India, Pakistan)
Malai Kulfi (India)
Mawa (Pakistan)
Muktagachar Monda (Bangladesh)
Orom (Mongolia)
Palpayasam (India, Nepal)
Pastillas De Leche (Philippines)
Peda (India)
Rabri (India)
Rasogolla (India)
Shrikhand (India)
Shrikhand Wadi (India)
Sikarni (Nepal)
Yema (Philippines)

Latin America

Cajeta (Latin America)
Chongos Zamoranos (Mexico)
Cola De Mono (Chile)
Dulce De Leche (Latin America)
Jamoncillo (Mexico)
Manjar Blanco (Latin Amercia)
Penus (Guyana)

Near East

Ayran (Turkey)
Hogut (Qatar)
Kanafeh Bil Jibn (Lebanon)
Keshkeh (Syria)
Kish/Kushk (Egypt, Lebanon)
Shenglish/Sorke (Syria)
Trachanas (Cyprus)

Source : FAO Animal Production and Health Paper No. 85, 1990 : *The Technology of Traditional Milk Products in Developing Countries*, Food and Agriculture Organisation of the United Nations, Rome.

TABLE 2

Average Milk Yield Per Animal Per Day in India, 1992-98

Year	*Cow (kg/day)*		*Buffalo (kg/day)*
	Desi	*Crossbred*	
1992-93	1.58	5.58	3.46
1993-94	1.66	5.65	3.57
1994-95	1.69	5.81	3.65
1996-97	1.80	6.14	3.91
1997-98	1.84	6.16	3.94
1998-99	1.78	6.46	3.91

Source : State Department of Animal Husbandry and Veterinary Services etc.

TABLE 3

Production of Compound Animal Feed by ClFMA Members, 1964-2001

('000 tonnes)

Year	*Cattle feed*	*Poultry feed*	*Other feeds*	*Total*
1964	25.0	14.4		39.4
1974	275.4	164.6		440.0
1984	750.0	406.7		1,157.2
1990	1,324.5	833.7		2,161.2
1994	1,446.2	1,074.6	18.9	2,539.7
1995	1,512.9	1,267.8	29.9	2,810.6
1999-2000	1,278.7	1,600.7	23.6	2,903.0
2000-01	1,240.0	1,519.3	33.9	2,794.1

Source : The Compound Livestock Feed Manufactures' Association of India (CLFMA).

TABLE 4
Estimated Availability of Animal Feed Concentrates, 2000-01

(Million Tonnes)

Crop/Product Category	*Grains/ Seed*	*By-Product Solvent Extracted (SE) Meals/ Brans*
Oilseeds	18.70	13.09
Cottonseed	0.79	0.63
Total Oilseed Meals		13.72
Wheat	66.55	5.48
Rice	88.50	6.84
Pulses	11.72	0.35
Total Brans/chunnies		12.67
Total Non-conventional Feeds	12.22	12.22
Total Grains	14.00	14.00
Total Damaged Grains	3.00	3.00
Total concentrate Feeds		55.61

Source : National Livestock Policy Perspective, Report of the Steering Group, Government of India, 1996.

TABLE 5
Availability of Dry Fodder, 2000-01

Category	*Crop*	*Grain Output (MT)*	*Crop Residue (MT)*	*Grain/ Residue Ratio*
(1)	*(2)*	*(3)*	*(4)*	*(5)*
Straws	Rice	88.50	102.66	1:1.20
	Wheat	68.55	89.12	1:1.30
	Barley	1.89	2.46	1:1.30
Stovers	Maize	11.53	46.12	1:4.00
	Small millet	0.66	2.64	1:4.00
	Sorghum	7.38	29.20	1:4.00

(Contd.)

TABLE 5 (*Contd.*)

(1)	(2)	(3)	(4)	(5)
	Pearl millet	6.18	24.72	1:4.00
	Ragi	2.72	10.88	1:4.00
Pulses tops	Tur	2.48	0.00	1:4.00
	Gram	3.87	15.48	1:4.00
	Other pulses	5.37	21.48	1:4.00
Sugarcane tops	Cane	301.44 (million bales)	75.36	1:0.25
Oilseeds	Groundnut	6.64	13.28	1:2.00
	Sesame	5.3	10.60	1:2.00
	Nigerseed	0.17	0.34	1:2.00
	Rapeseed/Mustard	4.26	8.52	1:2.00
	Saflower	0.19	0.38	1:2.00
	Sunflower	0.71	1.42	1:2.00
	Soyabean	4.89	0.00	0.00
	Linseed	0.26	0.00	1:2.00
Total			453.66	

Source : Estimates by Dr. M.P.G. Kurup, Livestock Conslultant.

TABLE 6

Livestock Feed and Fodder Availability in 1993 and 2000 and Projection for 2006

(Million tonnes)

Category	*1993*		*2000*		*2006*	
	Demand/ Supply	*% Deficit*	*Demand/ Supply*	*% Deficit*	*Demand/ Supply*	*% Deficit*
Dry Fodder	583.62/ 398.88	31.00	632.61/ 523.61	17.30	561.80/ 483.26	13.98
Green Fodder	744.73/ 573.50	23.00	830.12/ 573.50	31.00	839.07/ 743.46	11.48
Concentras	79.40/ 41.98	47.10	88.05/ 46.18	47.60	127.09 45.63	64.10

Sources : 1. Report of the Policy Advisory Group on Integrated Grazing Policy, 1993, Ministry of Environment and Forests, Government of India.

2. Sub-Group on Feed Production Enhancement for the 10th Five Year Plan, 2001.

TABLE 7

Market Share of Veterinary Pharmaceuticals by Categories, 2001

(Percentage)

Category	*Market Share*	*Growth*
Antibiotics	22.0	29.0
Feed Additives & supplements	18.5	13.0
Tonics	18.9	30.0
Anthelmintics	8.0	19.0
Biological	6.8	40.0
Antibacterial	4.0	15.0
Anticoccidals	1.5	–5.0
Ectoparasiticides	2.7	48.0
Total Market		Rs. 11,000 million

TABLE 8

Prices of Milk and Milk Products, 1960–2001

Year	*Toned Milk (Rs./Litre)*	*Ghee (Rs./Kg.)*
1960	1.20	12.00
1974	2.50	25.00
1980	2.90	40.00
1985	3.60	50.00
1991	7.00	95.00
1994	8.00	120.00
1996	11.00	125.00
2001	14.00	125.00

Source : *Ibid.*

Bibliography

Acharya, R.M. (ed.) (1983) : Dairy India 1983, *Dairy India Year Book*, New Delhi.

———, (1984) : "Constraints is Milk Production Enrichment in India", *Diamond Justice Souvenir*, (1983-84), NDRI, Karnal.

Ant, G.D., D.S. Sidhu and S.S. Gill (1974) : "Milk Production Function and Ludhiana District", *Indian Journal of Dairy Science*, 27(4) : 284-89.

Benerjee, A. (1996) : "Indian Dairys : An Overview", *Productivity*, Vol. 36, No. 4.

———, (1983) : "Operation Flood : Its Achievements in South India", *Tamil Nadu Journal of Co-operation*, 74(889) 479-81.

———, (1999) : "Export Potential of Indian Dairy Products", *Indian Dairyman*, Vol. 50, No. 4.

Betra, J.D. (1979) : "Dairy Co-operated for Acceleratory Rural Development", *Kurukshetra*, January.

Chandhari, A.K. (1969) : "Farming for Milk", *Indian Dairyman*, Vol. 21, No. 7383.

Chrumilam, F. (1983) : "Importance of Animal Energy in the Indian Economy", *Khadigramodyog*, 29(10) 355, 63.

Croofy, R. (1980) : "Cattle, Economics—Development", Oxford IBH Haston Company, New Delhi.

Clinic Heavery (1951) : "Milk Production and Control", N.K. Lewis and Co., London.

Clarence, H.E. (1951) : "Milk and Milk Products", McGraw Hill, New York.

Chandan, R.C. (1997) : "Dairy-based Ingredients", Eagan Pren. and Paul Minnerate, USA, p. 17.

Davis, R.F. (1965) : "Modern Dairy Cattle Management", *Prentice Hall of India*, New Delhi.

Das, A.C. (1991) : "Operation Flood and Indian Dairying", Sage Publications, New Delhi.

De, Sukumar (2005) : "Outline of Dairy Technology", Oxford University Press, New Delhi.

———, (1980) : Outline of Dairy Technology, Ghee, Oxford University, Press, New Delhi.

Davis, W.L. (1933) : Indian Indigenous Milk Products, Thecker Spink and Co., Calcutta.

Elango, R. *et. al.* (1981) : Cost Analysis of Dairying, *Eastern Economists*, 77(1), 422-23.

Eckles, *et. al.* (2004) : Milk and Milk Products, Tata McGraw Hill Publications, New Delhi.

Everett, L. *et. al.* (1948) : Dairy Manufacturing Process.

Fox, P.F. *et. al.* (1998) : Dairy Chemistry, New York.

George, S. (1985) : Operation Flood : An Appraisal of Current Indian Dairy Policy, Oxford University Press, Delhi.

Hoddy, E. (1986) : Indian White Revolution, *Indian Dairyman*, 38(10), 491-92.

Huria, V. *et. al.* (1980) : Dairy Development in India, Some Critical Issues, *Economics and Political Weekly*, 15(45-46), 1931-42.

Heady, E.O. (1964) : Milk Production Function in Cooperating Various for Cow Characteristics and Environments, *Journal of Farm Economics*, 46(1) 1-19.

I.B.A. (1999) : Dairy Industry: IBA Bombay.

John, P. (1975) : Economics of Dairy Development in India, Prahlad Prakashan, Patna.

John, V.H. (1983) : Operation Flood and the Rural Poor; Institute of Development Studies, Sunsex, U.K.

———, (1990) : Operation Flood and Indian Dairy, Sage Publications, New Delhi.

Krishan, R.C. (1979) : Dairy and Milk, *Yojna*.

Katju, Vailash Nath (1966): "Dairy Farming", *Khadigramodyog*.

Khurody, D.N. (1974) : Dairying in India : A Review, Asia Publication House, Bombay.

Kanawjia, S.K. (2000) : "Technical Advances in Paneer Making", *Indian Dairyman*, 52(10), 45–50.

Lalwani, M. (1987) : Effects of Technological Change in Dairy Farming in India, Delhi School of Economics.

Mohanan, N. (1972) : Milk Co-operatives in India.

Mishra, S.N. (1979) : Livestock Planning in India : Vikas Publishing House, New Delhi.

Nair, K.N. (1985) : "White Revolution in India : Facts and Issues", *EPW*, 20 (25 and 26).

Nair, K.N. (1979) : Alternative to Operation Flood-II Strategy, *EPW*, 16(52), 21-29.

NDDB (1994) : Women's Dairy Co-operatives Leadership Programme, NDDB, Anand.

Patel, *et.al.* (1988) : "Shrikhand : A Review", *Indian Journal of Dairy Science.*

Patric John, (1975) : Economics of Dairy Development in India, Patna.

Singh, R.K.P. and A.K. Choudhary (1999) : Dairy Development through Co-operative, Speelbound Publishing, Rohtak.

Singh, R.P. and J.K. Singh (1994) : Dynamics of Dairy Development in Bihar, Indian Dairyman.

Shah, T.V. Vallabh *et. al.* (1995) : Institutional Structures for Dairy Development : India's Post-Independence Experience, *Indian Dairyman*, Vol. 47, No. 1, pp. 1-96.

Sharma, V.P. and S.K. Dutta (1999) : Economic Impact of WTO Agreement in Indian Dairy Industry, *Indian Dairyman*, Vol. 50, No. 11, pp. 15-35.

Govt. of India, Union Budgets, 2001-02 to 2006-07.

Govt. of India, *Economic Survey*, 1991-92 to 2005-06.

Govt. of Bihar, State Annual Budget, 2006-07, Patna.

Govt. of Bihar (2006), White Paper, 2006.

Various Issues of

Kurukshetra.
Yojana.
Economics and Political Weekly.
Indian Dairyman.
Annual Report, COMPFED.
The Economic Times.
The Financial Express.
The Business Standard.
The Times of India.
The Hindustan Times.
The Indian Express.
The Hindu.

Index